2nd Edition

Linear Algebra

Sun Tae Soh

Department of Mathematics
Myong Ji University

KYOWOO

Kyowoo Co., Ltd.
8, Yangnyeongsi-ro, Dongdaemun-gu,
Seoul, Korea 130-861,

Phone: 82-2-925-2861
F a x: 82-2-925-2860
e-mail: kyowoo@kyowoo.co.kr
http://www.kyowoo.co.kr

Printed in the Republic of Korea.
ISBN 979-11-251-0282-3 (93410)

₩ 23,000

Dedicated to the memory of my teacher, late Prof. George R. Kempf.

Contents

Preface to the Second Edition

This book, *Linear Algebra*, was first published in 2005 and then was improved, as time goes, in the following impressions, up to 2015. Since more time has passed after that, with the fixing of some still-remaining typos and errors of mine as much as I could, I added in this second edition a little bit of new materials to the latest impression of my book. Thus, this edition would be called the final version of my book, *Linear Algebra*.

As a remark of this preface, I would like to mention briefly what I had foreseen in 1999, 20 years ago, what will be the main aspect of Mathematics in the 21st century, that is included as an Appendix A: An Introduction to Algebraic Analysis,[1] of my book, because most of the people around the world nowadays recognize the importance of Computation Related Human Activities(CRHA), after watching AlphaGo's victory in a Go-game competition in Seoul, in the Spring of 2016. Since the main aspect of CRHA consists of (1) Symbolic Computation, i.e., Algebraic Analysis, (2) deductive reasoning or inductive reasoning for intellect's understanding, and (3) decision making process in order to reach a meaningful conclusion after finitely many steps in reasoning process.

I think it is very important for us, human-beings, to understand what each of these three actually means, and it is fairly well described in the Appendix A: An Introduction to Algebraic Analysis, of my book, although I did not mention them explicitly in terms of technical terminologies, because, as it is well-known that the Greek Philosophy of those philosophers such as Socrates (470 - 399 BC), Plato (428/427 or 424/423 – 348/347 BC), and Aristotle (384 - 322 BC) is based on the method of Mathematical proof, still nowadays only well-taught in the courses of Pure

[1]http://ch.catholic.or.kr/pundang/4/soh/mathnet_973071.htm

Mathematics, which had already well been developed by the Pythagorean School of 6th century BC, CRHA are also definitely based on the method of Mathematical proof, still nowadays only well-taught in the courses of Pure Mathematics.

I believe that people well know that there is no contradiction among any facts, as long as they are all Mathematically proved to be true, and hence this is the main reason why the method of the Mathematical proof is very important to us, human-beings, as long as we reason why. Another thing is that people usually do not make a mistake in deriving a true conclusion through a deductive reasoning from sound hypotheses, but people sometimes derive a false conclusion through an inductive reasoning even from sound hypotheses that is NOT a mathematical inductive reasoning. One should also recognize that Artificial Intelligence is born to have such a capacity but it inevitably has, like us human-beings, a defect in its inductive reasoning process.

I think the Appendix A: An Introduction to Algebraic Analysis and the first chapter of my book is in fact well-suited for the beginners who want to learn how to better think and understand CRHA in the Computer Age, the 21st century, because their main underlying theme is not about any particular subjects in Mathematics, but rather about how we should think and reason to derive a meaningful conclusion in an axiomatic manner.

Toward this direction, some amount of my research works and papers on finding or establishing a new relationship among objects, i.e., concepts/ideas, in Computation Related Human Activities(CRHA), that are meaningful to us human-beings, existing generically or specially among the given ideas/concepts, under the environment of internet-based Artificial Intelligence, which have been produced by the author for a period of more than recent 15 years, are available to the readers through the internet access.[2]

Sun T. Soh
March 2019

[2] http://ch.catholic.or.kr/pundang/4/soh/a_list_of_research_outputs.htm

The rest of the preface is the original preface to the first edition of this book:

Preface [to the First Edition]

We sometimes learn serious mathematics through Linear Algebra and this is partially because its significance later shows up almost everywhere in apparently totally separated parts of mathematics. Or, we may say that Linear Algebra is a backbone of quite many mathematical results discovered and proved until the end of 19th century. For instance, it is Linear Algebra through which we learn the very fundamental concepts such as those of Space and Time.

We recall that we extend our number system, the set $\mathbb{N} = \{1, 2, 3, ...\}$ of all natural numbers, to the set $\mathbb{Z}$ of all integers from our sincere efforts of obtaining, for instance, a solution for X for a given linear equation, $X + n = 0$ for a given number $n \in \mathbb{N}$. Here, we note that the word, natural, means "God-given".

This is perhaps our original source for the right definition of -3 as "a (new, I mean, non-natural) number whose addition to the given number 3 results in the 0, i.e., emptiness", i.e. $(-3) + 3 = 0$. Of course, we "know" that we have only one such a new number if ever it exists within our domain of thoughts. Thus, given a number 3, "*a linear relation of being equal to each other*", $(-3) + 3 = 0$, exactly tells us that we "know" either -3 from 0 or 0 from -3, as long as we live in $\mathbb{Z}$, i.e., we accept these (non-natural numbers) as new members of our numbers.

Similarly, we extend our number system, $\mathbb{Z}$, to the set $\mathbb{Q}$ of all rational numbers from our sincere efforts for obtaining, for instance, a solution for X for a given linear equation, $nX = 1$ for a given non-zero number $n \in \mathbb{Z}$. Similarly, this process thus allows us to have the right definition of the new number $\frac{1}{3}$. It is a number whose multiplication to the given number 3 results in 1, i.e., the unit. Of course, we have only one such a new number if ever it exists within our domain of thoughts. Thus, given a number 3, "*a linear relation of being equal to each other*", $\frac{1}{3} \cdot 3 = 1$, exactly tells us that we "know" either $\frac{1}{3}$ from 1 or 1 from $\frac{1}{3}$, as long as we live in $\mathbb{Q}$, i.e., we accept these (another non-natural numbers) as new members of our numbers.

This fundamental and conceptual process in Algebra also teaches us to easily conclude that $-(-1) = 1$, which only means that you should know how to prove this equality by yourself!

What are we then supposed to do when we are given a system of linear equations in several unknowns? We need to learn how to solve such a linear system and this is all about Linear Algebra. Thus Linear Algebra will tell you what's really going on when we attempt to solve a system of linear equations. For instance, once you learn this subject a little bit, you can tell others explicitly and affirmatively whether such a system has a solution or not, and if so, you can actually tell others about how to compute all the solutions of the system in a systematic manner. Understanding this part of Linear Algebra is very important because almost every subject in Linear Algebra can be expressed as a system of linear equations.

Then what about a system of non-linear, i.e.., polynomial equations in several unknowns? Our understanding is that virtually no books on Linear Algebra attempt to mention this, and consequently undergraduate students who take Linear Algebra courses never have a chance of thinking about it in advance. To break this old and out-dated tradition apart, we added some Appendix materials to this book which should be good enough to introduce the readers to this very interesting and important subject. However, since it requires some amount of knowledge on polynomials in several variables, where multiplication between two unknowns is common, it is the next subject to study if you are, for instance, an Algebra-oriented person. One should note that this brand new computational and algorithmic subject started its way only in the middle of 1960's.

This book has grown out of my teachings on the subject for several years in my University. In writing this book, I took M. Artin's famous book, *Algebra*, as a basic model.

In the main treatment of this book, we try to cover all the basic materials of Linear Algebra, particularly under linear, i.e., no multiplication between two unknowns is allowed, computational aspects of mathematics as much as we could. Here are about the contents of each chapter of this book.

In Chapter One, we introduce the most basic terminology in Algebra in general with some explanation for motivation. Since the materials are treated in an axiomatic manner, it is sometimes helpful for the beginners to have a chance of looking at the bare skeleton of the whole subject as roughly as possible, and this was the main purpose of preparing our chapter one.

Thus, Chapter 2 is the real beginning of this book. Like other books in Linear Algebra, we first review basic operations on matrices, and we directly move to the requirement of invariance of the solution set of a system of linear equations under our arithmetic treatments of equations. Here, we are talking about the fundamental concept of the three elementary row operations and their associated elementary matrices. Since it is useful in inductively proving certain properties, block multiplication of matrices is also discussed. In particular, the determinant of a square matrix is first axiomatically defined, and then it is explained that it can be defined in several different ways by its uniqueness property, which is the direct result of this axiomatic definition of the determinant.

In Chapter 3, we first discuss the basic concepts of vector spaces such as linear independence, span, basis and dimension. Then we discuss the effect of change of basis. Our treatment of change of basis is very well prepared. As an application, we discuss row space, column space and null space of a matrix. This will give students another chance of reviewing elementary row operations.

In Chapter 4, we first study linear transformations and associated matrices. Then eigenvectors and eigenvalues of a linear operator are discussed. Since this is also about the effect of a good change of basis, we begin our talk on diagonalization problem of matrices, which is still the subject of the later chapter. We also briefly treat the so-called rigid motions in two or three dimensional spaces.

In Chapter 5, we discuss bilinear forms in general. This subject is usually studied as one studies the so-called quadratic forms, but in our geometry-oriented axiomatic approach of this chapter quadratic forms are treated as computable examples; hence in this sense our treatment is quite abstract.

In Chapter 6, we develop a kind of factorization in an advanced Linear Algebra, based on generators and relations among them of finitely generated R-modules. The main computing means of this chapter is the diagonalization process of integer matrices. In particular, we prove the structure theorem for abelian groups. We also discuss rational canonical forms and Jordan canonical forms in this chapter.

Moreover, we also intend this book to be considered as an introduction to the subject, called **Algebraic Analysis**, that is, symbolic computation. Unless we approximate numbers which is very common in Numerical Analysis, the nature of all the objects under our mathematical consideration are exact. However, unlike those materials in Linear Algebra, we need to multiply two unknowns and this requires us to study a polynomial ring in several (i.e., more than one) variables over an (algebraically closed) field, called "Computational Algebraic Geometry", which is beyond the scope of the main theme of this book. On the other hand, since even Personal Computers nowadays provide us with an excellent Hardware means for symbolic computation and since this theory provides us with a surprising means for solving a system of polynomial equations, we added some computed examples in each chapter which requires a Computer Algebra System, MAPLE, and these examples should convince the reader the importance of symbolic computation, or Algebraic Analysis.

As a Second Appendix, we mention a little bit of MAPLE commands for Linear Algebra. Since this book is not written for MAPLE, one of commercially available Computer Algebra System(CAS), our description is very brief. For more about Computer Algebra Systems, one may visit our homepage:

http://trinitas.mju.ac.kr/intro2softwares.html .

There, you will further find very useful informations for users on CAS, MAPLE, REDUCE, Matlab, etc.[3] Also,

http://trinitas.mju.ac.kr/intro2InetCompu.html

describes an important aspect of the availability of Internet environment to everybody in this wonderful world of 21st century.[4]

[3]http://ch.catholic.or.kr/pundang/archive/4/intro2softwares.html
[4]http://ch.catholic.or.kr/pundang/archive/4/intro2InetCompu.html

CHAPTER 1

A Few Fundamental Concepts and Definitions in Algebra

In this introductory chapter, we are going to review a few basic concepts and definitions in Algebra. Before we start what constitutes Algebra, we would like to have a look at what had happened in our numbers, first. The main purpose of this chapter is for the readers eventually to have a good sense in Symbolic Computation or Algebraic Analysis. Another reason is that quite many axioms used in defining algebraic objects are in fact those satisfied by natural numbers, integers, or rational numbers.

1. Natural Numbers and Their Extensions

Every human being knows what are natural numbers, otherwise he is not one of us. That is, we know that if we add the number one to the number one, then we have the number two, and if we add the number one to the number two, then we have the number three, etc., i.e., we can keep going on this indefinite process to obtain another natural number repeatedly. This is our first incidence of recognizing an infinite set, the set of all natural numbers,

$$\mathbb{N} = \{1, 2, 3, 4, ...\},$$

and nobody knows exactly why and how we have recognized natural numbers in this way.

Moreover, we "know" that every non-empty subset S of $\mathbb{N}$ has a least element in it, i.e., we "accept" this as a fact without its proof. We also "know" that **the associative laws**

$$a + (b + c) = (a + b) + c \qquad \text{and} \qquad a \cdot (b \cdot c) = (a \cdot b) \cdot c.$$

hold for both addition and multiplication of any three natural numbers a, b, and c. We also "know" that **the distributive laws**

$$a \cdot (b + c) = a \cdot b + a \cdot c \quad \text{and} \quad (a + b) \cdot c = a \cdot c + b \cdot c$$

hold for natural numbers, etc. This means that when we say that we "know" natural numbers, we in fact means that we also accept these axioms as facts, and we are ready to apply these axioms in our later arguments on numbers. Since everybody knows about what constitutes natural numbers these numbers are some times called God-given numbers. That is, "natural" means "God-given", and conversely.

Then how about the negative numbers? Suppose a wife has prepared a tonight's dinner for 3 family members, including herself. But, her husband brought 2 friends with him to have dinner together at his house tonight. Then, his wife will quickly figure out that there will be no food for her, and this will make her unhappy. The situation is abstracted as follows. Given the number 3, she basically figures out the relationship that

$$(-3) + 3 = 0,$$

where recognizing the zero, 0, or the negative number, -3, requires the recognition of the other. The point here is that she recognizes this relation first in her head; and then the usefulness of the negative number -3 and/or the number 0 (here meaning that nothing left to her) as a terminal of this relationship. Here, by definition,

DEFINITION 1.1. *0 is a number whose addition to any number n is any number n itself.*

Since this kind of arguments is always possible for any other number n, our slow abstraction leads us to the introduction of the set of integers,

$$\mathbb{Z} = \{..., -3, -2, -1, 0, 1, 2, 3, ...\}.$$

Thus, *by definition*, the number -3 is a number whose addition to the natural number 3 results in the number 0 (i.e., nothing left). We then easily check that if there were somewhere such a number -3 then it must be unique. But since this number is not an element of $\mathbb{N}$, we need to extend the set $\mathbb{N}$ of natural numbers to the set $\mathbb{Z}$ of integers, so that we can do such an arithmetic.

This can be more abstractly expressed as follows: Solve the following equation for the unknown X:

$$X + 3 = 0.$$

Since its solution -3 is not one of the natural numbers, we had to extend our number system from $\mathbb{N}$ to $\mathbb{Z}$ in order to satisfy arithmetic demands of our everyday life like the above example.

With this information, we can prove that

LEMMA 1.1.

$$-(-1) = 1.$$

PROOF. For, $-(-1)$ exactly means that it is a number whose addition to the number -1 results in the number 0 (i.e., nothing left). That is,

$$-(-1) + (-1) = 0.$$

But, we also know

$$1 + (-1) = 0.$$

Then since there is only one number whose addition to -1 results in the number 0, we conclude that $-(-1) = 1$. □

Similarly, suppose we are given a 12-inch round pizza and 4 persons are going to equally share it. Then we obviously know how to divide it into 4 pieces of equal size and equal shape, and this establishes an abstract relationship,

$$\frac{1}{4} \cdot 4 = 1,$$

which equivalently says that the addition of the same $\frac{1}{4}$ six times is equal to the number 1, or equivalently it says that $\frac{1}{4}$ is a number whose multiplication to the natural number 4 is the number 1. Thus, by definition $\frac{1}{4}$ is a number whose multiplication to 4 results in the number 1 (i.e., a unit.) Here, **by definition**, 1 is a number whose multiplication to any number n is any number n itself.

We then easily check that if such a number exists then it must be unique. Since this kind of arguments is always possible for any other number n, our slow abstraction leads us to the introduction of the set $\mathbb{Q}$ of rational numbers, which is an extension of $\mathbb{Z}$. This can be more abstractly expressed as follows: Solve the following equation for the unknown X:

$$X \cdot 4 = 1.$$

Since its solution $\frac{1}{4}$ is not one of the integers, we had to extend our number system from $\mathbb{Z}$ to $\mathbb{Q}$ in order to satisfy arithmetic division demands of our everyday life like the above example.

Moreover, we have

LEMMA 1.2.

$$0 \cdot 1 = 0.$$

PROOF. This is because, for any number n, we have

$$\begin{aligned} 0 \cdot 1 &= (n + (-n)) \cdot 1 \\ &= n \cdot 1 + (-n) \cdot 1 \\ &= n + (-n) \\ &= 0. \end{aligned}$$

□

On the other hand, Greek mathematicians thought that the set of positive rational numbers must be everything, i.e., there are no other numbers they can think of. But, this was a big mistake. For this, consider a rectangular triangle of two equal sides 1. Then even Babylonians in almost $4,000$ years ago knew (without proof) that there *is* a number whose square is equal to $1 + 1 = 2$, from the ruler and compass construction of magnitudes, i.e., positive numbers. This number is represented as the square root of 2:

$$\sqrt{2}.$$

LEMMA 1.3. *$\sqrt{2}$ is not a rational number.*

PROOF. Suppose that it is a rational number, say $\frac{a}{b}$ with $(a, b) = 1$, i.e., a and b are relatively prime. Then that whose square is two exactly means that $a^2 = 2b^2$; hence a^2 is divisible by 2; hence a itself is divisible by 2; hence $4a'^2 = 2b^2$. This just implies that b is again divisible by 2, which is but a contradiction to our choice of a and b being relatively prime to each other. □

This can be more abstractly expressed as follows: Solve the following equation for the unknown X:

$$X^2 = 2.$$

Obviously, the disciples of Pythagoras should know about this proof, but history tells that they were forbidden officially to discuss this strange number, $\sqrt{2}$.

Another famous example is π. By its geometric definition, it is a ratio of two magnitudes, the length of circumference of a circle to that of the diameter of the same circle. Thus, if both of these magnitudes are rationals then π must be a rational number. Thus, the question is: when we choose the diameter of our circle to be the unit 1, the length (i.e., the magnitude) of the circumference of the same circle is whether rational or not.

Of course the right answer to this question took a pretty long time and we nowadays know that π is not a rational number. The history of mathematics tells us that pretty some time ago one mathematician spent his whole life to see if the decimal representation of π is periodic or not. (It is a fact that if its decimal representation is periodic, then it is a rational number.) Of course, he could not prove that it is periodic during his life time since by dividing infinitely many times for decimal fractions, he attempted to prove what he must prove its complete opposite, and another mathematician eventually proved that it is not a rational number at all, spending something like almost more than 20 years of his life time!

We note, unlike other examples as we have seen so far, that π is not accepted as a number as a solution of an equation containing an unknown X.

In fact, it took more than 1,800 years since the Greek period for us to properly define the real numbers using the concept of limits, or limiting process. In fact, it turns out that among real numbers, there are more irrational numbers (i.e., non-rational numbers) than rational numbers.

Having accepting the real numbers $\mathbb{R}$ as numbers, we still have a difficulty if we are asked to solve an equation for X:

$$X^2 + 1 = 0.$$

Obviously, there are no real solutions to this equation, because a square of any number is positive unless it is zero. On the other hand, if there is a number whose square is -1, then mathematicians in Italy of Renaissance period figured out that a very interesting thing happens.

2. What is Algebra?

The subject Algebra has a long history in Mathematics. For instance, even in Greek period, people knew how to add, subtract, multiply, and divide the **magnitudes**, even if they did not know how to write down numbers efficiently and even if they did not have symbols for these four basic operations, $+, -, \times$, and $\div$. They did all their mathematical thinking in a verbal way, based on the so-called method of "Dialogues". Even with the Roman way of representing numbers people had a great difficulty in multiplying numbers. As time goes, people very slowly began to introduce several efficient symbols to represent numbers and the basic four operations on numbers, etc.

When the followings are what we are about to deal with, we call it **Algebra**.

- We treat both **numbers** such as $1, 2, 3, \ldots$, and **symbols** such as $a, b, c, \pi, x, y, z, \ldots$.

- Some of the symbols, for instance, $a, b, c, \ldots$ etc., are already **known** to us and others, for instance, $x, y, z, \ldots$etc., are yet **unknown** to us, but they are treated as if they were numbers, regardless of being known or unknown to us.
- These numbers and symbols are connected using the four connectives: $+, -, \cdot$, and $\div$ (**Addition, Subtraction, Multiplication, and Division**, respectively).
- We "**equate**" two such expressions with $=$ (Equality) sign to have an equation.
- We apply the following rules (or principles) to an equation:
  - (**Addition rule**) Whenever we add a number or a symbol on one side of an equality, we have to do the same thing on the other side of the equality. [Whenever we subtract a number or a symbol from one side of an equality, we have to do the same thing on the other side of the equality.]
  - (**Multiplication rule**) Whenever we multiply [divide] a non-zero number or a symbol on one side of an equality, we have to do the same thing on the other side of the equality.
  - (**Substitution rule**) Whenever we substitute a symbol in one side of an equality with another symbol or number, we have to do the same thing on the same symbol on the other side of the equality.
- We try to express any unknown symbol to be equal to a **polynomial expression** of the rest of known and unknown symbols and numbers.

REMARK 1.1. *We remark that the* ***decimal notation*** *of numbers such as* 12345 *and those* ***connecting symbols*** *such as* $+, -, \cdot, \div$, *and* $=$ *were not generally used almost until the end of* 15th *century. Before this time, most mathematical expressions are given as plain sentences, as are in Euclid's Elements. For a long time, the method of the so called "****Dialogues****" has been the best method of learning mathematics.*

3. What is Linear Algebra?

It is certainly a branch of Algebra with only one restriction: We **never** multiply unknown symbols or quantities together (which results in more unknown symbols). This means that linear algebra is about how to solve a system of linear equations, and it is about how to use such a computational method in handling other parts of mathematics as applications. Whenever this restriction is applied, it belongs to the subject, called **Linear Algebra**. Since computation itself has no taste at all, we usually need to understand what's really going on behind such computations. This means that we need to learn new concepts which enable us to study certain important properties and/or results in Linear Algebra.

On the other hand, the existence of a subject called Linear Algebra then also means the existence of a non-linear Algebra in general, where several unknown symbols or quantities are allowed to be multiplied together. For instance, starting 1965 we are now able to solve a system of polynomial equations, thanks to Austrian Mathematician Bruno Buchberger's discovery of an algorithm for computing the so-called "Groebner Basis" for a polynomial ideal in several variables. To learn more about this, students need to learn polynomial rings in one and/or several variables, which is certainly beyond the treatments of this course.

For more about this, you may refer to Appendix A of this book.

4. Several Basic Definitions in Algebra

Before we go into Linear Algebra, let's have a look at several basic definitions in Algebra. In the long run, this will make you very comfortable in learning Linear Algebra as well as Algebra.

DEFINITION 1.2. *A **field** F is an integral domain where every nonzero element a is invertible, i.e., for each $a \neq 0$ there exists b in F such that $ab = 1$.*

Such an a is called a **left inverse** of b and such a b is called a **right inverse** of a.

EXAMPLE 1.1. *$\mathbb{Q}$ = the set of all rational numbers, $\mathbb{R}$ = the set of all real numbers, and $\mathbb{C}$ = the set of all complex numbers.*

DEFINITION 1.3. *An **integral domain** D is a commutative ring where the **cancellation law** holds, i.e., if $ab = ac$ and $a \neq 0$ then $b = c$ in D.*

REMARK 1.2. *Thus, an integral domain is such a place where you can cancel out a non-zero element, even without multiplying (mostly non-existing in D) its inverse on both sides of the equality under consideration.*

EXAMPLE 1.2. *Any fields, $\mathbb{Z}$ = the set of all integers, and $\mathbb{Q}[X]$ = the set of all polynomials in one unknown X with rational coefficients.*

DEFINITION 1.4. *A **commutative ring** K is a ring whose multiplication is commutative.*

EXAMPLE 1.3. *Any integral domains and the quotient ring $\mathbb{Z}/n\mathbb{Z}$ where n is a positive integer.*

DEFINITION 1.5. *A **ring** R is an Abelian group together with multiplication $R \times R \to R$ satisfying the following rules:*

1. *(Existence of multiplicative identity) There exists 1 in R such that $r.1 = r = 1.r$ for any element r of R.*
2. *(Associativity) $a(bc) = (ab)c$.*
3. *(Distributive law) Multiplication is distributive over addition in both directions, i.e., $a(b + c) = ab + ac$ and $(a + b)c = ac + bc$.*

EXAMPLE 1.4. *The set of all $n \times n$ matrices, where $n = 1, 2, 3, \ldots$*

REMARK 1.3. *Thus, there are two binary operations in a ring and they are usually called "addition (denoted by $+$)" and "multiplication (denoted by $\cdot$)".*

DEFINITION 1.6. *An **Abelian group** A is a group whose binary operation, $+$, is commutative, i.e., $a + b = b + a$.*

REMARK 1.4. *Its "additive identity" is usually denoted by 0.*

DEFINITION 1.7. *A **group** G is a set together with a binary operation $*$ satisfying the following rules:*

(1) *(Existence of identity) There exists an element e in G such that $g * e = g = e * g$ for each element g of G.*
(2) *(Associativity) $a * (b * c) = (a * b) * c$.*
(3) *(Invertibility) Every element of G is invertible.*

EXAMPLE 1.5. $\mathbb{Z}$.

REMARK 1.5. *We "**know**" a **set** S only when we know each of its elements. When have a finite set, we have no problem at all. But, most of the sets we are going to deal with in mathematics are infinite sets. For example,*

$$\mathbb{N}= \textit{the set of all natural numbers } 1, 2, 3, ...$$

*is the most important infinite set whose infiniteness we all know. We recall that the traditional meaning of English word "natural" is "God-given". Natural numbers are also called "**counting** numbers" or "**ordering** numbers" for an obvious reason that they are always used in counting and/or ordering objects.*

REMARK 1.6. *If you read the above definitions backward, you will be able to read them in a proper logical order in defining them.*

Thus, there could be no binary operation on a set, we have one binary operation on a group G, and we have two binary operations $+$ and $\cdot$ on a ring R.

Since the main subject of this book is Linear Algebra, we need to define additional ones:

DEFINITION 1.8. *A **vector space** V over a field F is a module over the commutative ring F.*

EXAMPLE 1.6. $\mathbb{R}^2$ = *the real plane, and* $\mathbb{R}^3$ = *the Euclidean 3-space.*

DEFINITION 1.9. *A (left) **module** M over a commutative ring K is an Abelian group together with an operation (called scalar multiplication):*

$. : K \times M \to M$ *satisfying the following rules: for each* a, b *in* K *and for each* m, n *in* M,

(1) $1.m = m$
(2) $(ab).m = a.(b.m)$
(3) $(a + b).m = a.m + b.m$
(4) $a.(m + n) = a.m + a.n$

EXAMPLE 1.7. *Any commutative ring* K *is a module over itself.*

EXAMPLE 1.8. *Any abelian group* A *is a module over* $\mathbb{Z}$.

DEFINITION 1.10. *An element* a *of a commutative ring* R *is a* ***zero divisor*** *if* $a \neq 0$ *and there exists* $b \neq 0$ *such that their product* $ab = 0$.

LEMMA 1.4. *A commutative ring* R *is an integral domain if and only if* R *has no zero divisors.*

PROOF. Suppose that R is an integral domain, that is, it satisfies the cancellation law. If $ab = 0$ and $a \neq 0$, then since

$$ab = 0 = a \cdot 0$$

holds, we cancel out a from both sides to conclude that $b = 0$,Thus, there are no non-zero b satisfying $ab = 0$ where $a \neq 0$. Conversely, if it has no zero divisors, it exactly means that $ab = 0$ implies either $a = 0$ or $b = 0$. Suppose that $ab = ac$ and $a \neq 0$. Then we have

$$a \cdot (b - c) = 0.$$

Then since $a \neq 0$ by hypothesis, $b - c$ must be zero, i.e., $b = c$, which is all about the cancellation law. □

DEFINITION 1.11. *A* ***binary operation*** *on a set* S *is a map,* $S \times S \longrightarrow S$*, from their Cartesian product* $S \times S$ *to* S.

We use maps to compare sets:

THEOREM 1.1. *Let* $f\colon X \longrightarrow Y$ *be a map.*

(1) f *is one-to-one, i.e., injective, if and only if it has a left inverse.*

(2) *f is onto, i.e., surjective, if and only if it has a right inverse.*

PROOF. Suppose that f is injective. For any y in the image of f we have only one preimage x, so we define this y to be sent to this x. For any y in Y outside the image of f, we define this y to be sent to the same particular element x_0 of X of our choice. This defines a map $g: Y \longrightarrow X$ such that their composition $g \circ f$ is the identity map: $X \longrightarrow X$. This proves the $\Rightarrow$ direction of the first assertion. The converse of the first assertion is obvious.

Now suppose that f is surjective. Then we note that the domain X of the function f is partitioned into the disjoint union of various non-empty preimages of the elements $y \in Y$. Thus, by the Axiom of the choice function, which is equivalent to the Axiom of choice, we can choose an element out of each preimage of elements of Y under the map f, and this defines a function $g: Y \longrightarrow X$ such that $f \circ g$ is the identity map: $Y \longrightarrow Y$. This proves the $\Rightarrow$ direction of the second assertion. The converse of the second assertion is also obvious. □

REMARK 1.7. *In particular, we note that our proof also tells us that such inverses are not uniquely determined.*

COROLLARY 1.1. *A map f is bijective if and only if it has both a right inverse and a left inverse. When this is the case, any right inverse of f is equal to any left-inverse of f.* □

COROLLARY 1.2. *A map f is bijective if and only if it has a (two-sided) inverse f^{-1}. When this is the case, any two inverses of f are the same, and*

$$(f^{-1})^{-1} = f. \quad \square$$

DEFINITION 1.12. *A map ϕ from a module M to a module N over the same commutative ring K is called a **linear** map or linear transformation if for each $m, m' \in M$ and for each $\alpha \in K$,*

$$\begin{aligned} \phi(m + m') &= \phi(m) + \phi(m'), \\ \phi(\alpha m) &= \alpha\phi(m), \end{aligned}$$

hold.

DEFINITION 1.13. *An **isomorphism** of two modules M and N is a linear map from M to N that is also one-to-one and onto.*

The readers are strongly recommended to memorize the above definitions in the given order.

There is one theorem which is really essential to us.

THEOREM 1.2. ***(Mathematical Induction)** A set of natural numbers that contains the number 1 and the number $n+1$ whenever it contains a number n must be the set of all numbers.* □

5. A Few Remarks on Connectives

Finally, we would like to discuss how the logical connectives are used. In particular, we will discuss the right way of using implication in mathematics. Suppose we are given two input statements p and q. Then the truth value of $p \wedge q$, $p \vee q$ and $p \Rightarrow q$ as outputs are defined as follows:

p	q	$p \wedge q$	$p \vee q$	$p \Rightarrow q$
T	T	T	T	T
T	F	F	T	F
F	T	F	T	T
F	F	F	F	T

Thus, we know immediately the truth value of the outputs $p \wedge q$ and/or $p \vee q$, once we know whether each of the two inputs p and q are true or false.

But the situation is quite different when the usage of the $p \Rightarrow q$ is involved. Although it is defined as is in the above table, its usage in mathematics is the following:

p	$p \Rightarrow q$	q
T	T	T
T	F	F
F	T	T
F	T	F

Here, the only input is p and we want to verify (or prove) whether q is true or false as an output, through a mathematical proof, which corresponds to the step $p \Rightarrow q$. When our input (or hypothesis) p is true, since the truth value of both $p \Rightarrow q$ and q are exactly the same by definition, our final result (or output) q must be true (false respectively), as long as our argument, $p \Rightarrow q$, is true (false respectively).

p	$p \Rightarrow q$	q
T	T	T
T	F	F

On the other hand, when our input (or hypothesis) p is false, our output q can be either true or false regardless of our argument $p \Rightarrow q$ being true. For instance, any mathematical conclusion from a false hypothesis, like 3 divides 5, could be either true or false regardless of our argument $p \Rightarrow q$ being true.

p	$p \Rightarrow q$	q
F	T	T
F	T	F

This is what it really says: Suppose you want to know whether a statement q is true or false. If you start from a false statement p and, via a completely error-free $p \Rightarrow q$ step, arrive at q, then, even though your argument or reasoning step $p \Rightarrow q$ is completely error-free, the very statement q is still true or false, which was but the initial state of the statement q!; hence you still do not know whether the statement q is true or false!

Now, here is our explanation on the so-called method of contradiction in our proof business and it goes as follows: Suppose we are able to reach a conclusion q through a true argument, i.e., a valid $p \Rightarrow q$ step, from a hypothesis p whose truth value is not yet verified, but on the other hand everybody is able to see that "that" conclusion q is in fact false, then the **only** conclusion we can draw is that the hypothesis p is false. This is in fact the exact meaning of the following table, where the truth value of p is the output.

p	$p \Rightarrow q$	q
F	T	F

EXAMPLE 1.9. *A good example of a proof by the method of contradiction is the proof of Lemma 1.3 above.*

6. Introductory Remarks on MAPLE Sessions

The followings are a brief explanation of some of the MAPLE commands and operations, including MAPLE-System interfaces.

REMARK 1.8. *Here are for system managers of Unix or Linux servers:*

(1) If a system manager of Unix or Linux server wants to add some MAPLE commands for all user's defaults, write them in the file, called

init

in the subdirectory "lib" of the MAPLE directory. (For example, /usr/local/maple/lib/init).

(2) If a system manager of Unix or Linux server wants to let MAPLE users have a message, write them in the file, called

message2init

in the subdirectory "lib" of the MAPLE directory. (For example, /usr/local/maple/lib/message2init).

REMARK 1.9. *The following list is for general interface between MAPLE and a user:*

(1) If a user wants to add some MAPLE commands for user's own defaults, write them in the file, called

.mapleinit

in the case of Unix or Linux user in the user's home directory.(For Unix or Linux Users) The following line sets up x11 graphics environment for MAPLE's default plot display to have a fast response for remote clients:

plotsetup(x11);

(2) The following line let the MAPLE echo whenever you read from a file or write to a file.

interface(echo=2);

(3) To read in the output of a MAPLE command later, namely, after saving it to a file using MAPLE's command,

writeto("file_name");

as an input to your MAPLE session, type in at MAPLE's prompt:

interface(prettyprint=0);

where the default setting of prettyprint is 2.

(4) Note that MAPLE's command,

read "file_name";

reads in the contents of the ASCII file, file_name, you create using the

writeto "file_name";

command.

REMARK 1.10. *The following line selects different plotdevice other than the default one. Using ghostview command at unix % prompt, you can display the graph of sin(x). This requires an X-window emulator. We also recall that .pdf file can be read at unix prompt using the command, "acroread filename" .*

———————

 # interface(plotdevice=x11, plotoutput=myfile);
interface(plotdevice=ps, plotoutput=myfile);
*p1:=plot3d(sin(x)*cos(y),x=-2*Pi..2*Pi,y=-2*Pi..2*Pi):*
*p2:=plot3d(x*y,x=-2*Pi..2*Pi,y=-2*Pi..2*Pi):*
display({p1,p2});
 # For instance,

ghostview myfile (enter)

at unix % prompt will display the graph of sin(x).

———————

REMARK 1.11. *The following list tells us how to start a MAPLE session and Linear Algebra related computations:*

(1) To start a MAPLE session, type in at a system prompt (in the case of Unix or Linux machine):

maple (and then press the Enter Key).

(2) To finish a MAPLE session, type in at a MAPLE prompt:

quit; (and then press the Enter Key),

where ";" is definitely needed to execute any MAPLE command.

(3) The following usage of ":" postpones the execution of the "command" until MAPLE receives the terminator, ";" :

command:

(4) Multiplication, denoted by

$$*$$

in MAPLE, is commutative, unless you use a different symbol for a non-commutative multiplication of which the first example is multiplication of two matrices.

(5) To load a subpackage for Linear Algebra in to a MAPLE session, type in:

with(linalg);

(6) For a non-commutative multiplication, we must use both "&*" *and* "evalm" symbols together for such a multiplication. For example,

evalm(A&*B);

will compute the product of two matrices A and B.

(7) Unlike most programming languages, like, C, or FORTRAN, it supports an automatic multiple-precision. This functionality is really required for an exact computation at all.

(8) It understand the meaning of certain symbols commonly used in mathematics. For example, it recognizes sqrt(2) as a square root of 2 knowing that its square is 2. Also, it recognizes I (the capital I) as $\sqrt{-1}$ knowing that its square is -1.

(9) For Numerical computation, we need to assign the number of digits for floating point computation and the following command, for instance, will assign 50 digits for numerical computation,

Digits:=50;

where ";" is definitely needed to execute any MAPLE command. The default value of *Digits* is 10. We note that this adjustment of effective number of digits is not Hardware dependent.

(10) Finally, whenever you have a question on a MAPLE command, or you want to figure out the exact command name for a mathematical operation, type in:

?command_name;

Then an informative message will be provided by MAPLE system.

Exercises for Chapter 1

EXERCISE 1.1. *Prove that to say that the cancellation law holds in a commutative ring R is the same as to say that R has no zero-divisors, i.e., If $ab = 0$ for $a, b \in R$ then one of either a or b is zero.*

EXERCISE 1.2. *Can you define a field consisting of only 2 elements?*

EXERCISE 1.3. *Can you define a field consisting of only 3 elements?*

EXERCISE 1.4. *Can you define a field consisting of only 4 elements?*

EXERCISE 1.5. *We obviously know that every non-empty subset S of the set $\mathbb{N}$ of all natural numbers has the least element in it under the usual natural* **order** $>$ *between any two numbers. Does this also hold for the set $\mathbb{Z}$ of all integers?*

EXERCISE 1.6. *Can you see that every Abelian group is a $\mathbb{Z}$-module?*

EXERCISE 1.7. *Derive a formula for the sum: $S_5(n) = 1^5 + 2^5 + ... + n^5$, and prove that your formula holds for each $n \in \mathbb{N}$. (Hint: Consider the difference $(k+1)^6 - k^6$ when k runs through from 1 to n.)*

CHAPTER 2

Matrix Operations

Matrices play an important role in this book. They provides many concrete examples, and constitutes an important part of the theory. We first review and learn the basics of matrix manipulation. Since matrices are used in much of mathematics, the techniques you are going to learn in this chapter are sure to be useful elsewhere.

1. The Basic Operations

Let m, n be positive integers.

DEFINITION 2.1. *An $m \times n$* ***matrix*** *A is a collection of mn numbers arranged in a rectangular array:*

$$A = \begin{bmatrix} a_{11} & \dots & a_{1n} \\ \vdots & & \vdots \\ a_{m1} & \dots & a_{mn} \end{bmatrix} = (a_{ij}) \tag{2.1}$$

where the number of rows is m and the number of columns is n.The numbers a_{ij} in a matrix A are called ***matrix entries****. i is called the* ***row index****, and j is called the* ***column index****.*

EXAMPLE 2.1. *For example, $\begin{bmatrix} 1 & 2 & 3 \\ 0 & 3 & 4 \end{bmatrix}$ is a 2×3 matrix whose $a_{12} = 2$.*

Thus, a_{ij} is the entry appearing in the ith row and j^{th} column of the matrix $A = (a_{ij})$ where $1 \leq i \leq m$ and $1 \leq j \leq n$.

REMARK 2.1. *When we make a list of quite many numbers only in a horizontal manner, we will soon fall short in the space. To overcome this kind of difficulty, after listing the first* n *elements,* $a_{11}, ..., a_{1n}$*, we list the next* n *elements,* $a_{21}.,,,a_{2n}$ *in the second row, and we repeat this until we list the* m^{th} n *elements,* $a_{m1}, ..., a_{mn}$*, and this constitutes an* $m \times n$ *matrix. Because of this, we sometimes identify the set of all* $m \times n$ *matrices with the set* $\mathbb{R}^{mn}$ *of all* mn*-tuples when the* a_{ij} *are real numbers, for instance. Namely, we will soon learn that the set of all* $m \times n$ *matrices with real entries is a vector space over* $\mathbb{R}$*, which is exactly the same as the* mn*-space* $\mathbb{R}^{mn}$ *as a vector space.*

The one by one matrix $[a_{11}]$ is exactly the same as the number a_{11}, since we do not need the [ ] or ( ) in this particular case.

DEFINITION 2.2. *A* $1 \times n$ *matrix is called an* n*-dimensional* ***row vector,*** *where we usually drop the index* i *when* $i = 1$*, and we simply write a row vector as*

$$A = [a_1, ..., a_n], \qquad \text{or as} \qquad A = (a_1, ..., a_n). \tag{2.2}$$

Similarly, an $m \times 1$ *matrix is called an* m*-dimensional* ***column vector,***

$$B = \begin{bmatrix} b_1 \\ \vdots \\ b_m \end{bmatrix}. \tag{2.3}$$

DEFINITION 2.3. *The* ***transpose*** A^t *of a matrix* $A = (a_{ij})$ *is a matrix whose entry* $b_{ij} = a_{ji}$ *for each* i *and* j*.*

1.1. Addition of two matrices. As is explained in Remark 2.1, we have:

DEFINITION 2.4. ***Addition*** *of two matrices (of the same size) is vector addition:*

$$(a_{ij}) + (b_{ij}) = (c_{ij}), \tag{2.4}$$

where $c_{ij} = a_{ij} + b_{ij}$ *for each* i *and* j*.*

EXAMPLE 2.2. *Thus,*

$$\begin{bmatrix} 1 & 2 & 3 \\ 4 & 5 & 6 \end{bmatrix} + \begin{bmatrix} 7 & 8 & 9 \\ 10 & 11 & 12 \end{bmatrix} = \begin{bmatrix} 8 & 10 & 12 \\ 14 & 16 & 18 \end{bmatrix}.$$

The sum of two matrices A, B is defined only when they are both of the same shape or size, that is, when they are $m \times n$ matrices with the same m and n.

1.2. Scalar multiplication.

DEFINITION 2.5. ***Scalar multiplication*** *of a matrix* (a_{ij}) *by a number* α *is defined as with vectors:*

$$\alpha(a_{ij}) = (b_{ij}),$$

where $b_{ij} = \alpha a_{ij}$ *for each* i *and* j*.*

EXAMPLE 2.3. *Thus*

$$2\begin{bmatrix} 1 & 2 & 3 \\ 4 & 5 & 6 \end{bmatrix} = \begin{bmatrix} 2 & 4 & 6 \\ 8 & 10 & 12 \end{bmatrix}.$$

Since $[s] = s$, numbers s will also be referred to as **scalars**.

1.3. Matrix multiplication.

DEFINITION 2.6. ***The product*** $C = (c_{ij})$ *of two matrices* $A = (a_{ik})$ *of size* $l \times m$ *and* $B = (b_{kj})$ *of size* $m \times n$ *is defined as:*

$$c_{ij} = \sum_{k=1}^{m} a_{ik} b_{kj}, \tag{2.5}$$

for each $1 \le i \le l$ *and each* $1 \le j \le n$*.*

In particular, when A is a row vector of size m and B is a column vector of size m, their product is usually called the **dot product**, i.e., the **inner product**.

EXAMPLE 2.4.

$$\begin{bmatrix} 1 & 2 & 3 \\ 4 & 5 & 6 \end{bmatrix} \begin{bmatrix} 1 \\ 4 \\ 3 \end{bmatrix} = \begin{bmatrix} 18 \\ 42 \end{bmatrix}.$$

Thus, if we have one more row below the second row, $\begin{bmatrix} 4 & 5 & 6 \end{bmatrix}$ *then we will have one more row below* 42*, and if we have one more column after* $\begin{bmatrix} 1 & 4 & 3 \end{bmatrix}^t$*, then we will have one more column after* $\begin{bmatrix} 18 & 42 \end{bmatrix}^t$*, etc.*

REMARK 2.2. *In the* 19th *century matrix notation was introduced to write down a system of linear equations more efficiently. Suppose we are given a system of linear equations:*

$$a_{11}x_1 + \dots + a_{1n}x_n = b_1$$

$$a_{21}x_1 + \dots + a_{2n}x_n = b_2$$

$$\dots\dots\dots\dots\dots\dots$$

$$a_{m1}x_1 + \dots + a_{mn}x_n = b_m$$

where a_{ij} *are scalars, i.e., known numbers, and* x_j *are unknowns. We can write this very efficiently as a* ***matrix equation:***

$$AX = B, \tag{2.6}$$

where

$$A = (a_{ij}) = \begin{bmatrix} a_{11} & \dots & a_{1n} \\ a_{21} & \dots & a_{2n} \\ & \dots & \\ a_{m1} & \dots & a_{mn} \end{bmatrix}, X = (x_j) = \begin{bmatrix} x_1 \\ . \\ . \\ x_n \end{bmatrix},$$

$$\textit{and } B = (b_i) = \begin{bmatrix} b_1 \\ . \\ . \\ b_m \end{bmatrix}$$

.

This alternative way of representing a system of linear equations is very efficient especially when the indices m *or* n *are relatively large.*

EXAMPLE 2.5. *Thus, the matrix equation*

$$\begin{bmatrix} 1 & 2 & 3 \\ 4 & 5 & 6 \end{bmatrix} \begin{bmatrix} x \\ y \\ z \end{bmatrix} = \begin{bmatrix} 18 \\ 42 \end{bmatrix}$$

represents the following system of two linear equations in three unknowns $x, y,$ *and* z*:*

$$1x + 2y + 3z = 18$$
$$4x + 5y + 6z = 42$$

Various *identities* are satisfied by the matrix operations, such as the **distributive law**:

$$A(B + C) = AB + AC \quad \text{and} \quad (A + B)C = AC + BC \tag{2.7}$$

and the **associative law**:

$$A(BC) = (AB)C. \tag{2.8}$$

These two laws hold whenever the matrices involved have suitable sizes, so that their products are defined. Because of the associative law, we usually write ABC instead of either $A(BC)$ or $(AB)C$.

Scalar multiplication is compatible with matrix multiplication in the obvious sense:

$$c(AB) = (cA)B = A(cB) \tag{2.9}$$

The proofs of these identities are straightforward and not very interesting.

In contrast, the commutative law does not hold for matrix multiplication; that is,

$$AB \neq BA, \text{ in general.} \tag{2.10}$$

EXAMPLE 2.6.

$$\begin{bmatrix} 0 & 1 \\ 0 & 0 \end{bmatrix}\begin{bmatrix} 0 & 0 \\ 0 & 1 \end{bmatrix} = \begin{bmatrix} 0 & 1 \\ 0 & 0 \end{bmatrix},$$

but

$$\begin{bmatrix} 0 & 0 \\ 0 & 1 \end{bmatrix}\begin{bmatrix} 0 & 1 \\ 0 & 0 \end{bmatrix} = \begin{bmatrix} 0 & 0 \\ 0 & 0 \end{bmatrix},$$

and hence their product is not commutative.

We can multiply both sides of a matrix equation

$$B = C$$

on the left by a matrix A, to conclude that $AB = AC$, provided that the products AB and AC are defined. Similarly, if the products are defined, then we can conclude that $BA = CA$. But from these two, we can not generally derive, for example,

$$AB = CA$$

from $B = C$.

Several matrices have their names: A matrix all of whose entries are 0 is called a **zero matrix** and is also denoted by 0. A matrix A is called a **diagonal matrix** if its only non-zero entries are diagonal entries, where those entries a_{ii} of A are called diagonal. A square matrix, i.e., a matrix of $n \times n$, whose only non-zero entries are 1 in each diagonal position,

$$I_n = \begin{bmatrix} 1 & 0 & . & . & . & 0 \\ 0 & 1 & . & . & . & . \\ . & & & . & & . \\ . & & & & . & . \\ . & & & & . & . \\ 0 & . & . & . & . & 1 \end{bmatrix} \tag{2.11}$$

is called the $n \times n$ **identity matrix**. It behaves like 1 in multiplication.

Let A be a square $n \times n$ matrix. If there exists a matrix B such that

$$AB = I_n \quad \text{and} \quad BA = I_n, \tag{2.12}$$

then B is called an **inverse** of A and is denoted by A^{-1}:

$$A^{-1}A = I_n = AA^{-1}. \tag{2.13}$$

When A has an inverse, it is said to be an **invertible matrix**.

The matrix $A = \begin{bmatrix} 1 & 2 \\ 3 & 4 \end{bmatrix}$ is invertible since its inverse is

$$A^{-1} = \begin{bmatrix} -2 & 1 \\ \frac{3}{2} & -\frac{1}{2} \end{bmatrix}.$$

Check this by computing the products AA^{-1} and $A^{-1}A$.

LEMMA 2.1. *An inverse is unique if it exists at all. In other words, there can be only one inverse.*

PROOF. Let B, C be two inverses of the same matrix A. Then by the associative law, we have

$$C = CI = C(AB) = (CA)B = IB = B,$$

thus $B = C$. □

REMARK 2.3. *We will see later (in Proposition 2.5) that A is invertible if there is a matrix B such that either one of the two relations $AB = I_n$ or $BA = I_n$ holds, and that B is then the inverse of A. But since multiplication of matrices is not commutative, this fact is not obvious.*

PROPOSITION 2.1. *Let A, B be invertible square matrices. Then so is their product AB, and*

$$(AB)^{-1} = B^{-1}A^{-1}.$$

More generally, if $A_1, ..., A_m$ are invertible, then so is their product $A_1 \cdot \cdot \cdot A_m$, and its inverse is $A_m^{-1} \cdots A_1^{-1}$.

PROOF. We need to check that $B^{-1}A^{-1}$ is the inverse of AB:

$$ABB^{-1}A^{-1} = AIA^{-1} = AA^{-1} = I,$$

and similarly,

$$B^{-1}A^{-1}AB = ... = I.$$

The last assertion is proved by induction on m: When $m = 1$, the assertion is trivial. Next we assume that the assertion is true for $m = k$, and we need to show that the assertion also holds for $m = k + 1$. Let P denote the product $A_1 \cdots A_k$ of k invertible matrices of the same size and let A_{k+1} be another invertible matrix of the same size. Then PA_{k+1} is invertible and its inverse is $A_{k+1}^{-1}P^{-1} = A_{k+1}^{-1}A_k^{-1} \cdots A_1$. This shows that the assertion is true for $m = k + 1$. Therefore, the last assertion is true for each $m \in \mathbf{N} = \{1, 2, 3, ...\}$ by the theorem of mathematical induction. □

EXAMPLE 2.7. *(MAPLE) This example explains how to create matrices and how to multiply them. We use the MAPLE:*

———————

interface(echo=2);
with(linalg);
A:=matrix([[1,2],[3,4]]);

$$A := \begin{bmatrix} 1 & 2 \\ 3 & 4 \end{bmatrix}$$

inverse(A);

$$\begin{bmatrix} -2 & 1 \\ 3/2 & -1/2 \end{bmatrix}$$

det(A);

$$-2$$

B:=matrix(2,2,[w,x,y,z]):evalm(A+B);

$$\begin{bmatrix} 1+w & 2+x \\ 3+y & 4+z \end{bmatrix}$$

*evalm(A&*B);*

$$\begin{bmatrix} w+2y & x+2z \\ 3w+4y & 3x+4z \end{bmatrix}.$$

——————

EXAMPLE 2.8. *(MAPLE) This example explains how to have an access to the entries of a matrix:*

——————

interface(echo=2);
with(linalg);
M1:=matrix([[1,2,3,4],[a,b,c,d],[e,f,g,h],[w,x,y,z]]);

$$M1 := \begin{bmatrix} 1 & 2 & 3 & 4 \\ a & b & c & d \\ e & f & g & h \\ w & x & y & z \end{bmatrix}$$

M1[2,3];

$$c$$

submatrix(M1,1..2,3..4);

$$\begin{bmatrix} 3 & 4 \\ c & d \end{bmatrix}$$

subvector(M1,2,1..4);

$$[a, b, c, d]$$

row(M1,4);

$$[w, x, y, z]$$

col(M1,2);

$$[2, b, f, x]$$

——————

EXAMPLE 2.9. *(MAPLE) This example shows how to perform elementary arithmetic operations on matrices:*

——————

interface(echo=2);
with(linalg);

M1:=matrix([[1,2],[3,4]]): transpose(M1);

$$\begin{bmatrix} 1 & 2 \\ 3 & 4 \end{bmatrix}, \begin{bmatrix} 1 & 3 \\ 2 & 4 \end{bmatrix}$$

evalm(1/M1);

$$\begin{bmatrix} -2 & 1 \\ 3/2 & -1/2 \end{bmatrix}$$

map(sqrt, M1);

$$\begin{bmatrix} 1 & \sqrt{2} \\ \sqrt{3} & 2 \end{bmatrix}$$

*evalm(M1^3), evalm(M1&*M1&*M1);*

$$\begin{bmatrix} 37 & 54 \\ 81 & 118 \end{bmatrix}, \begin{bmatrix} 37 & 54 \\ 81 & 118 \end{bmatrix}$$

———————

1.4. Block Multiplication. Various tricks simplify matrix multiplication in certain cases. **Block multiplication** is just one of them. Let N, M be $m \times n$ and $n \times p$ matrices, and let r be a positive integer less than n. We may decompose the two matrices into *blocks* as follows:

$$N = \begin{bmatrix} A & B \end{bmatrix} \quad \textit{and} \quad M = \begin{bmatrix} C \\ D \end{bmatrix}, \tag{2.14}$$

where A has r columns and C has r rows. Then the matrix product can be computed as follows:

$$NM = AC + BD. \tag{2.15}$$

EXAMPLE 2.10. *Thus, we can compute:*

$$\begin{bmatrix} 1 & 0 & 2 \\ 0 & 1 & 3 \end{bmatrix} \begin{bmatrix} 4 & 5 \\ 6 & 7 \\ 0 & 0 \end{bmatrix} = I \begin{bmatrix} 4 & 5 \\ 6 & 7 \end{bmatrix} + \begin{bmatrix} 2 \\ 3 \end{bmatrix} 0 = \begin{bmatrix} 4 & 5 \\ 6 & 7 \end{bmatrix}.$$

We note that the formula (2.15) looks the same as the dot product of two vectors of same size.

We may also multiply matrices which are divided into more blocks. A decomposition into 4 blocks will be the most useful in an inductive proof of later results in this book. In this case, the rule for block multiplication is the same as for multiplication of 2×2 matrices: Let $r + s = n$ and let $k+l = m$. Assume that two matrices M of size $m \times n$ and N of size $n \times p$ are decomposed into submatrices

$$M = \begin{bmatrix} A & B \\ C & D \end{bmatrix} \quad \text{and} \quad N = \begin{bmatrix} A' & B' \\ C' & D' \end{bmatrix},$$

where the number of columns of the submatrix A is the same as that of the rows of A', for instance. Then similar to multiplication rule for 2×2 matrices, we have the rule for block multiplication:

$$\begin{bmatrix} A & B \\ C & D \end{bmatrix} \begin{bmatrix} A' & B' \\ C' & D' \end{bmatrix} = \begin{bmatrix} AA' + BC' & AB' + BD' \\ CA' + DC' & CB' + DD' \end{bmatrix}.$$

Again, this rule can be checked directly from the definition of matrix multiplication. In general, block multiplication can be used whenever two matrices are decomposed into submatrices in such a way that the needed intermediate matrix products are defined.

Not only facilitating computations, block multiplication is also a useful tool for proving facts about matrices by mathematical induction.

Exercises for Section 2.1

EXERCISE 2.1. *What are the entries a_{21} and a_{23} of the matrix*

$$\begin{bmatrix} 1 & 2 & 3 \\ 6 & 5 & 4 \\ 7 & 8 & 9 \end{bmatrix}?$$

EXERCISE 2.2. *Compute the product AB and BA for the following values of A and B:*

$$(1)\ A = \begin{bmatrix} 1 & 2 & 3 \\ 6 & 5 & 4 \end{bmatrix},\ B = \begin{bmatrix} -5 & -3 \\ 3 & 6 \\ 4 & 7 \end{bmatrix},$$

(2) $A = \begin{bmatrix} 4 & 2 \\ 1 & 3 \end{bmatrix}$, $B = \begin{bmatrix} 2 & -3 \\ 4 & 5 \end{bmatrix}$,

(3) $A = \begin{bmatrix} 1 \\ -1 \\ 0 \end{bmatrix}$, $B = \begin{bmatrix} 2 & 1 & 3 \end{bmatrix}$.

EXERCISE 2.3. *Verify the associative law for the matrix product for the following product:*

$$\begin{bmatrix} 1 & 3 \\ 0 & 2 \end{bmatrix} \begin{bmatrix} 0 & 1 & 2 \\ 3 & 1 & 2 \end{bmatrix} \begin{bmatrix} 1 \\ 3 \\ 7 \end{bmatrix}.$$

EXERCISE 2.4. *Compute* $\begin{bmatrix} 1 & 1 \\ 0 & 1 \end{bmatrix}^n$ *for* $n = 2, 3, 4$.

EXERCISE 2.5. *Find a formula for* $\begin{bmatrix} 1 & 1 \\ 0 & 1 \end{bmatrix}^n$ *for a general* n*, and prove it by induction on* n.

EXERCISE 2.6. *Compute the following matrix products by block multiplication:*

$$\left[\begin{array}{c|cc} 0 & 1 & 3 \\ \hline 0 & 1 & 0 \\ 2 & 0 & 2 \end{array}\right] \left[\begin{array}{c|cc} 1 & 3 & 4 \\ \hline 3 & 1 & 2 \\ 4 & 0 & 3 \end{array}\right].$$

EXERCISE 2.7. *We call an* $n \times n$ *matrix* $A = (a_{ij})$ ***upper triangular*** *if entries* $a_{ij} = 0$ *for* $i > j$*.Prove that the product of two upper triangular matrices is again upper triangular.*

EXERCISE 2.8. *Prove that the product to two diagonal matrices of the same size is commutative.*

EXERCISE 2.9. *Prove that a matrix which has a row of zero is not invertible.*

EXERCISE 2.10. *The* ***trace*** *tr of a square matrix is the sum of its diagonal entries. Prove the followings:*

a: $tr(A+B) = tr(A) + tr(B), \ tr(AB) = tr(BA)$.
b: *If B is invertible, then* $tr(A) = tr(BAB^{-1})$.

2. Row Reduction

Suppose we are given a system of linear equations:

$$a_{11}x_1 + ... + a_{1n}x_n = b_1$$
$$a_{21}x_1 + ... + a_{2n}x_n = b_2$$
$$.................$$
$$a_{m1}x_1 + ... + a_{mn}x_n = b_m$$

We can easily see that:

(1) Swapping any two of these equations does not change its solution set.
(2) Multiplying a non-zero number to both sides of any one equation does not change its solution set.
(3) Multiplying a non-zero number to both sides of an equation and then adding the resulting equation vertically to another equation does not change its solution set.

In other words, the solution set of the original system of linear equations is **invariant** under any finite combination of these three manipulation on equations. Thus, applying this important observations, which reflect our point of view in properly doing Linear Algebra discussed in Chapter 1, we try to rewrite the given system of linear equations into another simpler one in the sense that we can easily figure out its solution set, possibly including the empty set.

Changing our notations, we can express this observation exactly in terms of matrices: A given system of linear equations corresponds to:

$$AX = B,$$

and the three manipulations of equations above correspond to the following three **elementary row operations**:

(1) (Type I) Multiply (row i) by a non-zero scalar c.
(2) (Type II) Replace the ith row X_i by $X_i + aX_j$, or add $a \cdot$ (row j) to (row i).
(3) (Type III) Interchange (row i) and (row j).

In terms of matrix multiplication, we then easily verify that these three elementary row operations correspond to *left multiplication* to both sides of the equality in $AX = B$ of the following **elementary matrices**:

(1) (Type I)

$$mulrow(i, c) = \begin{bmatrix} 1 & & & & & & \\ & \cdot & & & & & \\ & & 1 & & & & \\ & & & c & & & \\ & & & & 1 & & \\ & & & & & \cdot & \\ & & & & & & 1 \end{bmatrix} \tag{2.16}$$

(2) (Type II)

$$addrow(j, i, a) \quad = \begin{bmatrix} 1 & & & & & & \\ & \cdot & & & & & \\ & & 1 & & & & \\ & & & \cdot & & & \\ & & & & \cdot & & \\ & & a & & & \cdot & \\ & & & & & & 1 \end{bmatrix} \tag{2.17}$$

$$or \quad \begin{bmatrix} 1 & & & & & & \\ & \cdot & & & a & & \\ & & \cdot & & & & \\ & & & \cdot & & & \\ & & & & 1 & & \\ & & & & & \cdot & \\ & & & & & & 1 \end{bmatrix} \tag{2.18}$$

(3) (Type III)

$$\text{swaprow}(i,j) = \begin{bmatrix} 1 & & & & & & \\ & \cdot & & & & & \\ & & 0 & & 1 & & \\ & & & \cdot & & & \\ & & 1 & & 0 & & \\ & & & & & \cdot & \\ & & & & & & 1 \end{bmatrix} \tag{2.19}$$

EXAMPLE 2.11. *The elementary* 2×2 *matrices are*

$$(i) \begin{bmatrix} c & 0 \\ 0 & 1 \end{bmatrix}, \begin{bmatrix} 1 & 0 \\ 0 & c \end{bmatrix}, \quad (ii) \begin{bmatrix} 1 & 0 \\ a & 1 \end{bmatrix}, \begin{bmatrix} 1 & a \\ 0 & 1 \end{bmatrix}, \quad (iii) \begin{bmatrix} 0 & 1 \\ 1 & 0 \end{bmatrix},$$

where a *and* c *are, as above, arbitrary non-zero numbers.*

We note that

LEMMA 2.2. *Elementary matrices are invertible, and their inverses are also elementary matrices.*

PROOF. The inverse of $swaprow(i,j)$ is itself, that of $mulrow(i,c)$ is $mulrow(i,1/c)$, and that of $addrow(j,i,a)$ is $addrow(j,i,-a)$. □

Successively applying these elementary row operations on a matrix A, we can **reduce** it to a simpler matrix $A' = E_k \cdots E_1 A$, where the E_i are any one of elementary matrices, as the following example shows. Such process is called **row reduction**, or **Gaussian elimination**.

EXAMPLE 2.12. *Suppose we want to solve a system of linear equations,* $AX = B$, *where*

$$A = \begin{bmatrix} 1 & 0 & 1 \\ 1 & 1 & 2 \\ 1 & 2 & 8 \end{bmatrix}, \quad X = \begin{bmatrix} x \\ y \\ z \end{bmatrix}, \quad \text{and } B = \begin{bmatrix} 5 \\ 7 \\ 12 \end{bmatrix}.$$

Combining these two matrices together, we obtain an augmented matrix M*:*

$$M = \begin{bmatrix} 1 & 0 & 1 & 5 \\ 1 & 1 & 2 & 7 \\ 1 & 2 & 8 & 12 \end{bmatrix}.$$

We then apply $addrow(1, 2, -1)$ *and* $addrow(1, 3, -1)$ *successively to obtain*

$$\begin{bmatrix} 1 & 0 & 1 & 5 \\ 1 & 1 & 2 & 7 \\ 1 & 2 & 8 & 12 \end{bmatrix} \to \begin{bmatrix} 1 & 0 & 1 & 5 \\ 0 & 1 & 1 & 2 \\ 0 & 2 & 7 & 7 \end{bmatrix}.$$

Then we apply $addrow(2, 3, -2)$ *to obtain*

$$\begin{bmatrix} 1 & 0 & 1 & 5 \\ 0 & 1 & 1 & 2 \\ 0 & 2 & 7 & 7 \end{bmatrix} \to \begin{bmatrix} 1 & 0 & 1 & 5 \\ 0 & 1 & 1 & 2 \\ 0 & 0 & 5 & 3 \end{bmatrix}.$$

Finally, we apply $mulrow(3, 1/5)$*,* $addrow(3, 1, -1)$ *and* $addrow(3, 2, -1)$ *successively to obtain*

$$\begin{bmatrix} 1 & 0 & 1 & 5 \\ 0 & 1 & 1 & 2 \\ 0 & 0 & 5 & 3 \end{bmatrix} \to \begin{bmatrix} 1 & 0 & 0 & 22/5 \\ 0 & 1 & 0 & 7/5 \\ 0 & 0 & 1 & 3/5 \end{bmatrix}.$$

From the last matrix, we can rewrite it as $A'X = B'$*, and then we easily see that* $x = 22/5, y = 7/5$*, and* $z = 3/5$*. That is, by successively applying elementary row operations, we are able to solve a system of linear equations in a systematic manner!*

EXERCISE 2.11. *Show that* $swaprow(i, j)$ *is redundant, i.e., not essential, by expressing* $swaprow(i, j)$ *in terms of* $addrow(j, i, a)$ *and* $mulrow(i, c)$*.*

The key observation is:

PROPOSITION 2.2. *The solutions of* $A'X = B'$ *are the same as those of* $AX = B$ *, where the augmented matrix* $M' = [A' \mid B']$ *is obtained from* $M = [A \mid B]$ *by applying row operations successively.*

PROOF. Write $M' = E_k \cdots E_1 M$. Then since each elementary matrix E_i is invertible, so is their product $P = E_k \cdots E_1$. Thus, if X is a solution of the original system $AX = B$, then $PAX = PB$, which is to say, $A'X = B'$, and hence X is also a solution of the new system. Similarly, the converse is obvious since P is invertible. $\square$

We now go back to row reduction of an arbitrary matrix. It is easy to see that, by a sequence of row operation, any matrix A can be reduced to a matrix which looks like:

$$\begin{bmatrix} 1 & *..* & 0 & *..* & 0 & *..* & 0 & & & \\ & & 1 & *..* & 0 & *..* & 0 & & & \\ & & & & 1 & *..* & 0 & . & . & . \\ & & & & & & 1 & & & \\ & & & & & & & . & & \\ & & & & & & & & . & \\ & & & & & & & & & . \end{bmatrix}, \tag{2.20}$$

where $*$ denotes any number and the large blank space in the lower part consists of all zeros. A matrix of this type is called a **row reduced echelon matrix**. The formal definition of a row echelon matrix is

DEFINITION 2.7. *A matrix is called* ***row reduced echelon*** *if the following three conditions are satisfied:*

(1) *The first non-zero entry in every row is* 1. *(This entry is called a* **pivot**.*)*
(2) *The first non-zero entry of row $i+1$ is to the right of the first non-zero entry of row i.*
(3) *The entries above a pivot are zero.*

REMARK 2.4. *A matrix is called in a* ***row echelon form*** *if the first two conditions of this definition is satisfied.*

EXERCISE 2.12. *The row-reduced echelon matrix obtained from a given matrix A by row reduction is unique, that is, it is not dependent on the* *particular sequence of operations applied. [Hint: Consider a system of linear equations $AX = 0$, which is always solvable.]*

LEMMA 2.3. *Let A be a square row reduced echelon matrix. Then either A is the identity matrix, or its bottom row consists of all zeros.*

PROOF. This is a direct consequence of the above definition. □

We have already seen the reason why row reduction is very useful in solving a system of linear equations. Moreover, a row reduced echelon matrix is useful in judging whether the given system is solvable i.e., **compatible**, or not:

PROPOSITION 2.3. *Let $M' = [A' \mid B']$ be a row reduced echelon matrix. Then the system of equations $A'X = B'$ is solvable, i.e., has a solution, if and only if there is no pivot in the last column B'. In that case, an arbitrary value can be assigned to the unknown x_i if column i does not contain a pivot.*

EXERCISE 2.13. *Prove this Proposition.*

We know that every system of homogeneous linear equations $AX = 0$ has the trivial solution $X = 0$. But looking at the row reduced echelon form again, we can conclude that such system has a *non-trivial* solution for X if there are more unknowns than the number of equations:

COROLLARY 2.1. *Every system $AX = 0$ of m homogeneous equations in n unknowns, with $m < n$, has a solution $X = \begin{bmatrix} x_1 \\ \vdots \\ x_n \end{bmatrix}$ in which some x_i is non-zero.*

PROOF. Let $A'X = 0$ be the associated row reduced echelon equation, and let r be the number of pivots. Then we automatically have $r \leq m$, and hence by the Proposition we may assign arbitrary values to $n - r$ variables. □

Moreover, if $X \neq 0$ is a solution of $AX = 0$ then its scalar multiple αX is also its solution for any scalar $\alpha \in \mathbb{R}$, and consequently there are infinitely many non-zero solutions of the homogeneous system.

EXAMPLE 2.13. *(MAPLE) This example shows how to perform elementary row operations step by step:*

———————

```
interface(echo=2);
with(linalg):
a:=matrix([[1,0,0,0,1,0,0,0],[5,2,0,0,0,1,0,0], \
[8,6,3,0,0,0,1,0],[0,9,7,4,0,0,0,1]]);
a1:=addrow(a,1,2,-5);
a2:=addrow(a1,1,3,-8);
a3:=addrow(a2,2,3,-3);
a4:=addrow(a3,2,4,-9/2);
a5:=mulrow(a4,2,1/2);
a6:=mulrow(a5,3,1/3);
a7:=addrow(a6,3,4,-7);
a8:=mulrow(a7,4,1/4);
aa:=submatrix(a,1..4,1..4);
b:=submatrix(a8,1..4,5..8);
check:=evalm(aa &* b);
b1:=vector([2,3,4,5]);
result:=linsolve(aa,b1);
check2:=evalm(aa &* result);
gausselim(a);
gausselim(aa);
```

———————

REMARK 2.5. *If you send those lines in the above example to*

maple@trinitas.mju.ac.kr

as a message of your E-mail of plain text format, written them between

input:

and

end input:

in the main body of your E-mail, then you will receive the computed results immediately, even with a "timex" stamp by a System.

This is about our ***InetCompu Service****, provided by the author of this book. This service allows a general public to use our High Performance Server, SGI Origin 3400, through our middleware for such an Internet Computing Service. For more about this, visit*

http://trinitas.mju.ac.kr/intro2InetCompu.html

or

http://trinitas.mju.ac.kr/intro2InetMaple.html

You will find a detailed information how to use our InetCompu Service, once you visit these two Internet Addresses.

Exercises for Section 2.2

EXERCISE 2.14. *Can you determine the elementary matrices corresponding to each elementary row operation applied in Example 2.12?*

EXERCISE 2.15. *Find all solutions of the system of linear equations* $AX = B$, *where*

$$A = \begin{bmatrix} 1 & 2 & 1 & 1 \\ 3 & 0 & 0 & 3 \\ 1 & -4 & -2 & -1 \end{bmatrix},$$

and $B = (0,0,0)^t$, $(1,1,0)^t$, *or* $(0,2,2)^t$.

EXERCISE 2.16. *What is the solution set of a single equation:* $x_1 + x_2 + 2x_3 - x_4 = 2$?

EXERCISE 2.17. *What are inverses of the following matrices?*

$$\begin{bmatrix} 2 & \\ & 2 \end{bmatrix}, \begin{bmatrix} 1 & \\ 1 & 1 \end{bmatrix}, \begin{bmatrix} 3 & 4 \\ 1 & 2 \end{bmatrix}.$$

EXERCISE 2.18. *Prove that if a product AB is invertible, then so are A and B.*

EXERCISE 2.19. *Prove that $(AB)^t = B^t A^t$ and $(A^t)^t = A$.*

3. Characterization of Invertible Matrices

We now use row reduction to characterize invertible matrices. In order to be invertible, a matrix must be square, first of all.

PROPOSITION 2.4. *Let A be a square matrix. The following statements are equivalent to each other:*

- **(a):** *A can be reduced to the identity matrix by a sequence of elementary row operations.*
- **(b):** *A is a product of elementary matrices.*
- **(c):** *A is invertible.*
- **(d):** *The system of homogeneous equations $AX = 0$ has only the trivial solution $X = 0$.*

PROOF. Each direction in $(a) \implies (b) \implies (c) \implies (d)$ can be easily proven, using the fact that each elementary matrix is invertible and Lemma 2.2. The direction $(a) \implies (d)$ is also very easy to prove.

To prove $(d) \implies (a)$, we suppose $\tilde{}(a)$, i.e., (a) does not hold for a matrix A. Then by Lemma 2.3 A is reduced by elementary row operations to a matrix A' whose last row is zero. Thus, $A'X = 0$ or equivalently $AX = 0$ has a non-trivial solution by Corollary 2.1, which is $\tilde{}(d)$. This completes the proof of Proposition 2.4. □

COROLLARY 2.2. *If a row of a square matrix is zero, then it is not invertible.* □

REMARK 2.6. *We shall see later that Proposition 2.4 grows as we study other subjects in Linear Algebra.*

Row reduction also provides a method of computing the inverse of an invertible matrix A: We reduce A to the (unique) row reduced echelon matrix

$$A' = E_k E_{k-1} \cdots E_1 A.$$

If $A' = I$ then by part (b) of Proposition 2.4 we see that $A = E_1^{-1} \cdots E_k^{-1}$ is invertible, and consequently we see that

$$A^{-1} = (E_1^{-1} \cdots E_k^{-1})^{-1} = E_k E_{k-1} \cdots E_1.$$

Otherwise, i.e., if $A' \neq I$, then a similar argument will show that A is not invertible. We thus proved:

COROLLARY 2.3. *Let A be a square matrix. To compute its inverse if ever possible, apply elementary row operations $E_1, ..., E_k$ to A, reducing to the row reduced echelon matrix A'. If A' is the identity matrix, then $A^{-1} = E_k E_{k-1} \cdots E_1$. Otherwise, A is not invertible.* □

The following example shows how we in practice apply this corollary.

EXAMPLE 2.14. *We want to see if the square matrix* $A = \begin{bmatrix} 1 & 0 & 1 \\ 1 & 1 & 2 \\ 1 & 2 & 8 \end{bmatrix}$ *is invertible and if so, we also want to compute its inverse. For this, we apply the same elementary row operations successively as we have done in Example 2.12 to the augmented matrix* $M = [A \mid I] = \begin{bmatrix} 1 & 0 & 1 & 1 & 0 & 0 \\ 1 & 1 & 2 & 0 & 1 & 0 \\ 1 & 2 & 8 & 0 & 0 & 1 \end{bmatrix}$ *to obtain* $M' = \begin{bmatrix} 1 & 0 & 0 & 4/5 & 2/5 & -1/5 \\ 0 & 1 & 0 & -6/5 & 7/5 & -1/5 \\ 0 & 0 & 1 & 1/5 & -2/5 & 1/5 \end{bmatrix}$. *We thus have shown that A is row reduced to the identity matrix. Therefore, A is invertible and its inverse is* $\begin{bmatrix} 4/5 & 2/5 & -1/5 \\ -6/5 & 7/5 & -1/5 \\ 1/5 & -2/5 & 1/5 \end{bmatrix}$.

We are also ready to prove

PROPOSITION 2.5. *Let A be a square matrix with either a left inverse B: $BA = I$, or a right inverse B: $AB = I$. Then A is invertible and B is its inverse.*

PROOF. Suppose that $AB = I$. We perform row reduction on A. By Lemma 2.3, there are elementary matrices $E_1, ..., E_k$ such that the row reduced echelon matrix $A' = E_k \cdots E_1 A$ of A is either the identity matrix or has bottom row zero. But, $A'B = E_k \cdots E_1$ is invertible, and hence its bottom row is not zero. From this it follows that A' has a non-zero bottom too. Thus, $A' = I$ and consequently $A = E_1^{-1} \cdots E_k^{-1}$is invertible by Proposition 2.4. But since $E_1^{-1} \cdots E_k^{-1} = (E_k \cdots E_1)^{-1}$ and the inverse of $(E_k \cdots E_1)^{-1}$ is $E_k \cdots E_1$, $A^{-1} = E_k \cdots E_1$.

For the other case, we simply replace the role of A and B in the above argument and conclude that B is invertible and A is its inverse. Thus, A is invertible too. □

EXAMPLE 2.15. *(MAPLE) The following lines compute the inverse of a square matrix aa:*

```
——————————
interface(echo=2);
with(linalg):
a:=matrix([[1,0,0,0,1,0,0,0],[5,2,0,0,0,1,0,0], \
[8,6,3,0,0,0,1,0],[0,9,7,4,0,0,0,1]]);
aa:=submatrix(a,1..4,1..4);
b:=inverse(aa);
check:=evalm(aa &* b);
——————————
```

Exercises for 2.3

EXERCISE 2.20. *Let A be a square matrix. Prove that there exists a set of elementary matrices, $E_1, ..., E_k$ such that $E_k \cdots E_1 A$ is either the identity matrix or has its bottom row zero.*

EXERCISE 2.21. *Let* $A = \begin{bmatrix} a & & & \\ 1 & a & & \\ & 1 & a & \\ & & 1 & a \end{bmatrix}$ *be a matrix where $a \neq 0$.*

(1) *Decide whether A is invertible.*

(2) *If so, compute its inverse.*

EXERCISE 2.22. *Do the same for* $A = \begin{bmatrix} 0 & 1 & 0 & 4 \\ 5 & 0 & 8 & 1 \\ 2 & 3 & 4 & 4 \\ 1 & 0 & 2 & 0 \end{bmatrix}$.

EXERCISE 2.23. *Prove that if A is invertible then $(A^{-1})^t = (A^t)^{-1}$.*

EXERCISE 2.24. *Prove that the inverse of a symmetric matrix is also symmetric.*

EXERCISE 2.25. *Let A and B be symmetric square matrices. Prove that the product AB is symmetric if and only if $AB = BA$.*

EXERCISE 2.26. *Prove that the reduced row echelon form obtained by row reduction of a matrix A is uniquely determined by A.*

4. Determinants

There are in fact several ways to define the function determinant $d : \mathbb{R}^{n\times n} \to \mathbb{R}$, where the domain consists of all $n \times n$ matrices for a given positive integer n, according to Remark 2.1. Here we are going to adopt an axiomatic approach in defining it. To have a better understanding of this approach we'd better to have a concrete motivation.

Motivation: Suppose that we are asked to solve a solution for the following differential equation with one condition:

$$\frac{df(x)}{dx} = f(x) \qquad \text{satisfying } f(0) = 1. \tag{2.21}$$

A theorem in Elementary Real Analysis will prove that the solution of this equation, if ever exists, is unique. On the other hand, (assuming that $f(x)$ is infinitely differentiable at $x = 0$) we apply *the method of power series*, to easily see that the function defined by

$$f(x) := 1 + \frac{1}{1!}x + \frac{1}{2!}x^2 + \cdots + \frac{1}{n!}x^n + \cdots,$$

for $-\infty < x < +\infty$ is infinitely differentiable at $x = 0$ and it is indeed a solution of (2.21). We denote this unique solution by e^x or $\exp(x)$.

Thus, we are going to abstractly define the determinant as a function $d : \mathbb{R}^{n\times n} \to \mathbb{R}$ satisfying certain finite number of conditions, and then we will present an explicit recursive function $\det(x)$ satisfying such conditions to show that our definition does make sense. Then we will derive several results to finally prove that there is only one such function.

DEFINITION 2.8. *A function $d : \mathbb{R}^{n\times n} \to \mathbb{R}$ is a* ***determinant function*** *if it satisfies the following 3 conditions:*

(1) $d(I) = 1$.
(2) *The function $d(A)$ is linear in the rows of the matrix A.*
(3) *If two adjacent rows of a matrix A are equal, then $d(A) = 0$.*

The second statement means that if $A = \begin{bmatrix} - R_1 - \\ \vdots \\ - R_n - \end{bmatrix}$, then by definition **linearity in the ith row** means that whenever R and S are row vectors, then

$$d\begin{bmatrix} \vdots \\ - R+S - \\ \vdots \end{bmatrix} = d\begin{bmatrix} \vdots \\ - R - \\ \vdots \end{bmatrix} + d\begin{bmatrix} \vdots \\ - S - \\ \vdots \end{bmatrix},$$

and

$$d\begin{bmatrix} \vdots \\ - cR - \\ \vdots \end{bmatrix} = cd\begin{bmatrix} \vdots \\ - R - \\ \vdots \end{bmatrix},$$

where c is a scalar and the other rows of the matrices appearing in these relations are the same throughout.

EXAMPLE 2.16. *Let* $\det : \mathbb{R}^{n\times n} \to \mathbb{R}$ *be a function defined recursively*

$$\begin{aligned} (2.22)\quad \det(A) &= a_{11}(-1)^{1+1}\det(A_{11}) + a_{21}(-1)^{2+1}\det(A_{21}) + \ldots \\ &\quad + a_{n1}(-1)^{n+1}\det(A_{n1}), \end{aligned}$$

where A_{ij} *denotes the* $(n-1)\times(n-1)$ *matrix obtained by crossing out the* i^{th} *row and the* j^{th} *column of* A *and* $\det[a]=a$ *for any one by one matrix* $[a]$. *For instance, when* $n=2$, *we have* $\det\begin{bmatrix} a & b \\ c & d \end{bmatrix} = a\det[d]-b\det[c] = ad-bc$.

Then (i) $\det(I)$ *is obviously equal to* 1, *(ii) by induction on* n, *one can easily prove that the recursive formula* $\det(A)$ *is linear in the rows of* A. *(iii) Let us now prove by induction on* n *that the* det *function satisfies the adjacency condition, i.e., the third condition in the above definition: Suppose that rows* j *and* $j+1$ *are equal. Then matrices* A_{i1} *also have two adjacent rows equal, except when* $i=j$ *or* $i=j+1$. *When* A_{i1} *has two adjacent equal rows, its determinant is zero by induction hypothesis. Thus, only two terms of the* ***minor expansion*** *(2.22) of* A *(through the* 1st *column) are different from zero, so we have*

$$\det(A) = a_{j1}(-1)^{j+1}\det(A_{j1}) + a_{j+1,1}(-1)^{j+2}\det(A_{j+1,1}).$$

Then since the two rows R_j *and* R_{j+1} *are equal,* $a_{j1}=a_{j+1,1}$ *and* $A_{j1} = A_{j+1,1}$, *and hence* $\det(A)=0$. □

EXAMPLE 2.17. *Here is what's really going on when a matrix* A *is explicitly given to us, say,* $A = \begin{bmatrix} 1 & 2 & 3 \\ 4+3 & 5+4 & 6+5 \\ 7 & 8 & 9 \end{bmatrix}$. *On the one hand, we have by definition,*

$$\begin{aligned} \det\begin{bmatrix} 1 & 2 & 3 \\ 4+3 & 5+4 & 6+5 \\ 7 & 8 & 9 \end{bmatrix} &= 1(-1)^{1+1}\det\begin{bmatrix} 5+4 & 6+5 \\ 8 & 9 \end{bmatrix} \\ &\quad +(4+3)(-1)^{2+1}\det\begin{bmatrix} 2 & 3 \\ 8 & 9 \end{bmatrix} \\ &\quad +7(-1)^{3+1}\det\begin{bmatrix} 2 & 3 \\ 5+4 & 6+5 \end{bmatrix}. \end{aligned}$$

On the other hand, we have by definition,

$$\begin{aligned}\det\begin{bmatrix}1&2&3\\4&5&6\\7&8&9\end{bmatrix} &= 1(-1)^{1+1}\det\begin{bmatrix}5&6\\8&9\end{bmatrix}\\ &\quad +4(-1)^{2+1}\det\begin{bmatrix}2&3\\8&9\end{bmatrix}\\ &\quad +7(-1)^{3+1}\det\begin{bmatrix}2&3\\5&6\end{bmatrix}\end{aligned}$$

and

$$\begin{aligned}\det\begin{bmatrix}1&2&3\\3&4&5\\7&8&9\end{bmatrix} &= 1(-1)^{1+1}\det\begin{bmatrix}4&5\\8&9\end{bmatrix}\\ &\quad +3(-1)^{2+1}\det\begin{bmatrix}2&3\\8&9\end{bmatrix}\\ &\quad +7(-1)^{3+1}\det\begin{bmatrix}2&3\\4&5\end{bmatrix}\end{aligned}$$

Thus, by adding RHS of these last two equalities under the induction hypothesis, we conclude that

$$\det\begin{bmatrix}1&2&3\\4+3&5+4&6+5\\7&8&9\end{bmatrix} = \det\begin{bmatrix}1&2&3\\4&5&6\\7&8&9\end{bmatrix} + \det\begin{bmatrix}1&2&3\\3&4&5\\7&8&9\end{bmatrix}.$$

Similarly, we conclude by induction on the size of the matrix that

$$\det\begin{bmatrix}1&2&3\\c\cdot 4&c\cdot 5&c\cdot 6\\7&8&9\end{bmatrix} = c\cdot\det\begin{bmatrix}1&2&3\\4&5&6\\7&8&9\end{bmatrix}.$$

LEMMA 2.4. *Let A be a square matrix.*

(a): *If a multiple of one row of A is added to an adjacent row, then the value of the determinant function d is unchanged.*

(b): *If two adjacent rows of A are interchanged, then the value of the determinant d is multiplied by -1.*

(c): *If two rows of A are equal, then $d(A) = 0$.*

PROOF. By linearity in rows and the third condition in the definition, (a) holds. To prove (b) we simply apply (a) repeatedly:

$$\begin{aligned} d\begin{bmatrix} R \\ S \end{bmatrix} &= d\begin{bmatrix} R \\ S-R \end{bmatrix} = d\begin{bmatrix} R+(S-R) \\ S-R \end{bmatrix} \\ &= d\begin{bmatrix} S \\ S-R \end{bmatrix} = d\begin{bmatrix} S \\ -R \end{bmatrix} = -d\begin{bmatrix} S \\ R \end{bmatrix}. \end{aligned}$$

To prove (c), we repeatedly apply (b) to A in order to obtain a square matrix A' whose two adjacent rows are the same. Then by (b)

$$d(A) = \pm d(A') = 0.$$

□

LEMMA 2.5. *Let A be a square matrix.*

(a'): *If a multiple of one row of A is added to another row, then the value of the determinant d is unchanged.*

(b'): *If two rows of A are interchanged, then the value of the determinant d is multiplied by -1.*

(c'): *If a row of A is zero, then $d(A) = 0$.*

PROOF. Since for any scalar α we have

$$\begin{aligned} d\begin{bmatrix} R \\ \vdots \\ \alpha R + S \end{bmatrix} &= d\begin{bmatrix} R \\ \vdots \\ \alpha R \end{bmatrix} + d\begin{bmatrix} R \\ \vdots \\ S \end{bmatrix} \\ &= \alpha d\begin{bmatrix} R \\ \vdots \\ R \end{bmatrix} + d\begin{bmatrix} R \\ \vdots \\ S \end{bmatrix} = d\begin{bmatrix} R \\ \vdots \\ S \end{bmatrix}, \end{aligned}$$

by parts (b) and (c) of the previous lemma, we obtain (a'). Since

$$
\begin{aligned}
0 &= d\begin{bmatrix} R+S \\ \vdots \\ R+S \end{bmatrix} \\
&= d\begin{bmatrix} R \\ \vdots \\ R \end{bmatrix} + d\begin{bmatrix} R \\ \vdots \\ S \end{bmatrix} + d\begin{bmatrix} S \\ \vdots \\ R \end{bmatrix} + d\begin{bmatrix} S \\ \vdots \\ S \end{bmatrix} \\
&= d\begin{bmatrix} R \\ \vdots \\ S \end{bmatrix} + d\begin{bmatrix} S \\ \vdots \\ R \end{bmatrix},
\end{aligned}
$$

we obtain (b'). If a row of A is zero, then $d(A)$ does not change when we multiply that row by 0. But by the second part of linearity condition in the definition, $d(A)$ gets multiplied by 0, Thus $d(A) = 0d(A) = 0$.This proves (c'). □

The linearity condition in the definition and properties (a') and (b') are about the effect of an elementary row operations on the value of the determinant function, and hence they can be rewritten in terms of the elementary matrices.

COROLLARY 2.4. *Let E be an elementary matrix in the previous Section, Row Reduction. Then*

$$
d(EA) = \begin{cases} cd(A) & \text{if } E \text{ is an elementary matrix of Type I} \\ d(A) & \text{if } E \text{ is an elementary matrix of Type II} \\ -d(A) & \text{if } E \text{ is an elementary matrix of Type III} \end{cases}.
$$

By putting $A = I$, we also obtain

COROLLARY 2.5. *The determinant of an elementary matrix is:*

(i): *(Type I) multiply a row by a non-zero constant c: $d(E) = c$.*
(ii): *(Type II) add a multiple of one row to another: $d(E) = 1$.*
(iii): *(Type III) row interchange: $d(E) = -1$.*

In particular, if we use these results cleverly, we get

LEMMA 2.6. *Let E be an elementary matrix and let A be a square matrix. Then*

$$d(EA) = d(E)d(A).$$

THEOREM 2.1. *(Axiomatic Characterization of the Determinant) The determinant function* det : $\mathbb{R}^{n\times n} \to \mathbb{R}$ *is the* only *one satisfying Definition 2.8.*

PROOF. Let $d : \mathbb{R}^{n\times n} \to \mathbb{R}$ be *any* function satisfying the defining properties in Definition 2.8. We have already seen that $\det : \mathbb{R}^{n\times n} \to \mathbb{R}$ satisfies these defining properties. We are going to prove that it must be the function $\det : \mathbb{R}^{n\times n} \to \mathbb{R}$.

For this, we first recall that any square matrix A can be reduced by elementary row operations to a matrix A' which is either the identity matrix or else its bottom row consists of all zeros:

$$A' = E_k \cdots E_1 A,$$

and hence by the above Lemmas and induction we see that

$$\left.\begin{matrix} 1 \\ 0 \end{matrix}\right\} = d(A') = d(E_k)d(E_{k-1}) \cdots d(E_1)d(A), \tag{2.23}$$

where $d(E_i)$ are determined as in the above corollary.

On the other hand, one can easily check that the function $\det : \mathbb{R}^{n\times n} \to \mathbb{R}$ which was defined in the above Example also satisfies the same value for elementary matrices E as the function d does. That is,

$$d(E) = \det(E)$$

for elementary matrices E. Thus, we also have $\det(EA) = \det(E)\det(A)$. Since the det function also satisfies the defining properties of the function d, we have

$$\left.\begin{matrix} 1 \\ 0 \end{matrix}\right\} = \det(A') = \det(E_k)\det(E_{k-1}) \cdots \det(E_1)\det(A). \tag{2.24}$$

Therefore, by equating these two expressions for $d(A')$ and $\det(A')$, and cancelling out $d(E_i)$ with $\det(E_i)$ from both sides, we conclude that $d(A) = \det(A)$.This completes the proof. □

REMARK 2.7. *This theorem also says the following: It does not matter whichever method is applied to construct the determinant function* $d : \mathbb{R}^{n\times n} \to \mathbb{R}$. *It exactly says that it is OK as long as it satisfies the defining properties in Definition 2.8, and it must be the same as the function* $\det : \mathbb{R}^{n\times n} \to \mathbb{R}$ *which was recursively defined in Example 2.16.*

NOTATION 2.1. *From now on, we are going to use the usual notation*

$$\det(A)$$

for the unique determinant of a square matrix A.

COROLLARY 2.6. *A square matrix* A *is invertible if and only if* $\det A \neq 0$.

PROOF. We interpret the expression in Eq. (2.24): The matrix A is invertible if and only if A' is the identity matrix if and only if $\det(A') = 1$ if and only if $\det(A) \neq 0$, because $\det(E_i) \neq 0$ for each elementary matrix E_i. □

We are now ready to prove one of the most important properties of the determinant function: its compatibility with matrix multiplication.

THEOREM 2.2. *Let* A, B *be any two* $n \times n$ *matrices. Then*

$$\det(AB) = \det(A) \cdot \det(B).$$

PROOF. We have already proved this when A is an elementary matrix (Lemma 2.6). For a general A we have two cases.

Case I: A is invertible. Then by part (b) of Proposition 2.4, A is a product of elementary matrices: $A = E_k \cdots E_1$. Thus, $\det(A) = \det(E_k) \cdots \det(E_1)$ by Lemma 2.6 and induction. Thus, $\det(AB) = \det(E_k \cdots E_1 B) = \det(E_k) \cdots \det(E_1) \cdot \det(B) = \det(A)\det(B)$.

Case II: A is not invertible. Then $\det(A) = 0$ by Eq. (2.24). On the other hand, its row reduced echelon form $A' = E_k \cdots E_1 A$ has bottom row zero, and hence the bottom row of $A'B$ is also zero; hence

$$0 = \det(A'B) = \det(E_k \cdots E_1 AB) = \det(E_k) \cdots \det(E_1) \cdot \det(AB).$$

Then since each $\det(E_i) \neq 0$, $\det(AB) = 0$. Therefore, both sides are equal as zero. □

COROLLARY 2.7. *If A is invertible, then* $\det(A^{-1}) = \frac{1}{\det(A)}$.

PROOF. $\det(A) \cdot \det(A^{-1}) = \det I = 1$. □

PROPOSITION 2.6. *Let A^t denote the transpose of A. Then*

$$\det A = \det A^t.$$

PROOF. If A is not invertible, then so is A^t; hence $\det(A) = 0 = \det(A^t)$. Otherwise, A is reduced to the identity matrix: $E_k \cdots E_1 A = I$; hence $A^t E_1^t \cdots E_k^t = I$. Then since $\det(E) = \det(E^t)$ for each elementary matrix, we easily conclude that $\det(A) = \det(A^t)$. □

COROLLARY 2.8. *The defining properties in Definition 2.8 for the* $\det(A)$; *hence their consequences also hold when we replace the word* row *with* column *throughout in this section.* □

EXAMPLE 2.18. *(MAPLE) The following lines compute the determinant of a square matrix:*

```
————————
interface(echo=2);
with(linalg):
a:=matrix([[1,0,0,0,1,0,0,0],[5,2,0,0,0,1,0,0], \
[8,6,3,0,0,0,1,0],[0,9,7,4,0,0,0,1]]);
aa:=submatrix(a,1..4,1..4);
det(aa);
b:=transpose(aa);
det(b);
————————
```

Exercises for 2.4

EXERCISE 2.27. *Evaluate the determinant of the following matrices:*

(1) $\begin{bmatrix} 1 & 2 & 3 \\ 6 & 5 & 4 \\ 7 & 8 & 9 \end{bmatrix}$

(2) $\begin{bmatrix} 1 & 5 & 8 & 0 \\ & 2 & 6 & 9 \\ & & 3 & 7 \\ & & & 4 \end{bmatrix}$

(3) $\begin{bmatrix} 1 & 3 & 5 & 7 \\ 0 & 2 & 4 & \\ 3 & 4 & & \\ 6 & & & \end{bmatrix}$

EXERCISE 2.28. *Compute the determinant of the following $n \times n$ matrix by induction on n:*

$$\begin{bmatrix} & & & & 1 \\ & & & 1 & \\ & & \cdot & & \\ & \cdot & & & \\ & 1 & & & \\ 1 & & & & \end{bmatrix}.$$

EXERCISE 2.29. *Prove that the Vandermonde determinant,*

$$\det \begin{bmatrix} 1 & 1 & 1 \\ a & b & c \\ a^2 & b^2 & c^2 \end{bmatrix} = (b-a)(c-a)(c-b).$$

EXERCISE 2.30. *Verify the claim made in part (ii) of Example 2.16.*

EXERCISE 2.31. *Prove that* $\det(-A) = (-1)^n \det(A)$, *where A is an $n \times n$ matrix.*

EXERCISE 2.32. *Let A and B be square matrices of the same size. Prove that* $\det(AB) = \det(BA)$.

EXERCISE 2.33. *Prove that* $\det \begin{bmatrix} A & B \\ 0 & D \end{bmatrix} = \det(A)\det(D)$, *where* A *and* D *are square blocks.*

EXERCISE 2.34. *Prove that*

$$\det \begin{bmatrix} 1 & 1 & \cdots & 1 \\ x_1 & x_2 & \cdots & x_n \\ x_1^2 & x_2^2 & \cdots & x_n^2 \\ \cdot & & & \cdot \\ \cdot & & & \cdot \\ \cdot & & & \cdot \\ x_1^{n-1} & & & x_n^{n-1} \end{bmatrix} = (-1)^{\frac{(n-1)n}{2}} \prod_{i<j}(x_i - x_j).$$

EXERCISE 2.35. *Prove that*

$$\det \begin{bmatrix} a+b+c & -c & -b \\ -c & a+b+c & -a \\ -b & -a & a+b+c \end{bmatrix} = 2(b+c)(c+a)(a+b).$$

EXERCISE 2.36. *Prove that*

$$\det \begin{bmatrix} a & b & b & b \\ a & b & a & a \\ a & a & b & a \\ b & b & b & a \end{bmatrix} = -(a-b)^4.$$

[Hint: $a = b$ *or not].*

5. Permutation Matrices

REMARK 2.8. *One of the reasons why we need to study permutation of matrices is that it will enable us to derive the formula, called the* complete expansion *of the determinant.*

$$\det A = \sum_{perm\ p} (sign\ p) a_{1p(1)} \cdots a_{np(n)}.$$

Since this formula has too many terms to be useful for computation unless n is small, it is not a good idea to discuss determinants first giving this formula as a definition of the determinant of a square matrix A.

This formula is rather theoretically important because we can consider determinants as polynomials in the n^2 variable matrix entries a_{ij}, with coefficients ± 1. Thus, for instance, if each matrix entry a_{ij} is a differentiable function of a single variable t, $a_{ij} = a_{ij}(t)$, then $\det(A)$ is also a differentiable function of t, because sums and products of differentiable functions are differentiable.

We start this section with a few definitions.

DEFINITION 2.9. *A bijective map p from a set S to itself is called a **permutation** of the set S: $p : S \longrightarrow S$.*

EXAMPLE 2.19. *Let $S = \{1, 2, 3\}$ and let $p(1) = 3, p(2) = 1$, and $p(3) = 2$. Then p is a permutation on the set $\mathbf{3} = \{1, 2, 3\}$. In particular, such a permutation is also called* cyclic, *because there is only one orbit of images under p.*

DEFINITION 2.10. *A **permutation matrix** P is a matrix with the following property: The operation of left multiplication by $P : A \longmapsto PA$ is a permutation of the rows of a matrix.*

EXAMPLE 2.20. *The elementary matrix of Type III (interchange of two rows) in Eq. (2.19) is the simplest permutation matrix. They correspond to the permutations called **transpositions**. Also,*

$$P = \begin{bmatrix} 0 & 1 & 0 \\ 0 & 0 & 1 \\ 1 & 0 & 0 \end{bmatrix}$$

is a permutation matrix sending a column vector

$$X = (x_1, x_2, x_3)^t \longmapsto PX = \begin{bmatrix} 0 & 1 & 0 \\ 0 & 0 & 1 \\ 1 & 0 & 0 \end{bmatrix} \begin{bmatrix} x_1 \\ x_2 \\ x_3 \end{bmatrix} = \begin{bmatrix} x_2 \\ x_3 \\ x_1 \end{bmatrix}.$$

P has permuted rows according to the cyclic permutation p given in the above Example: $1 \longmapsto 3 \longmapsto 2 \longmapsto 1$.

We read it as follows: The entry *of the* 1st *row is moved to the* 3rd *row, that of the* 3rd *row is moved to the* 2nd *row, and that of the* 2nd *row is moved to the* 1st *row. Or, we can read it as follows: The inverse* p^{-1} *of the permutation p is:* $1 \longmapsto 2 \longmapsto 3 \longmapsto 1$ *and the indices in* x_i *are moved according to* p^{-1}*, i.e.,* $\begin{bmatrix} x_1 \\ x_2 \\ x_3 \end{bmatrix} \longmapsto PX = \begin{bmatrix} x_2 \\ x_3 \\ x_1 \end{bmatrix} = \begin{bmatrix} x_{p^{-1}(1)} \\ x_{p^{-1}(2)} \\ x_{p^{-1}(3)} \end{bmatrix}$.

The permutation matrix P can be written conveniently either in terms of the **matrix unit** e_{ij}, i.e., a matrix whose entries are all zero except that its ij^{ih} entry is 1, or in terms of certain column vectors called the **standard basis** e_i, i.e., column vectors whose entries are all zero except that its ith position is 1.

From the above example we immediately conclude that

LEMMA 2.7. *The construction*

$$P = \left[e_{p(1)}, e_{p(2)}, ..., e_{p(n)}\right],$$

that is, the $p(j)^{th}$ *row of* P *is the* j^{th} *row of the identity matrix* I*, gives rise to a one-to-one correspondence between the set of all permutations on* $\{1, 2, ..., n\}$ *and the set of all permutation* $n \times n$ *matrices.* □

REMARK 2.9. *We note that there are two ways to associate a permutation to a permutation matrix P: the permutation p which describes how P permutes the* entries, *and the inverse operation which describes the effect on* indices. *They are related as follows: When we permute the* entries *of a vector* $(x_1, x_2, x_3)^t$ *according to a permutation p, the* indices *are permuted in the opposite way.* Our decision on this matter is that we will say that the permutation associated to a permutation matrix P is the one which describes its action on the **entries** of a column vector.

COROLLARY 2.9. *Let P be the permutation matrix associated to a permutation p.*

(a): *The* j^{th} *column of* P *is the column vector* $e_{p(j)}$.

(b): P *is a sum of* n *matrix units:* $P = e_{p(1)1} + \cdots + e_{p(n)n} = \sum_{j=1}^{n} e_{p(j)j}$. □

COROLLARY 2.10. *Let* P *be a square matrix. Then it is a permutation matrix if and only if it has a single* 1 *in each row and in each column, the rest of its entries being* 0. □

PROPOSITION 2.7.

(a): *Let* p, q *be two permutations, with associated permutation matrices* P, Q. *Then the matrix associated to the permutation* pq *is the product* PQ.

(b): *A permutation matrix* P *is invertible, and its inverse is the transpose matrix:* $P^{-1} = P^t$.*(That is, a permutation matrix is* orthogonal.*)*

PROOF. By pq we mean the composition of the two permutations

$$pq(i) = p(q(i)).$$

Since P operates by permuting rows according to p and Q operates by permuting according to q, the associative law for matrix multiplication tells us that PQ permutes according to pq:

$$(PQ)X = P(QX).$$

Thus, PQ is the permutation matrix associated to pq. This proves (a). For (b), the first assertion is obvious. For the second assertion, we go as follows: If $P = \left[e_{p(1)}, e_{p(2)}, ..., e_{p(n)}\right]$, then $P^t = \left[e^t_{p(1)}, e^t_{p(2)}, ..., e^t_{p(n)}\right]^t$, and hence their product

$$\left[e^t_{p(1)}, e^t_{p(2)}, ..., e^t_{p(n)}\right]^t \left[e_{p(1)}, e_{p(2)}, ..., e_{p(n)}\right] = [a_{ij}]$$

where

$$a_{ij} = e^t_{p(i)} e_{p(j)} = \begin{cases} 1 & \text{if } i = j \\ 0 & \text{otherwise} \end{cases}.$$

Thus, $P^{-1} = P^t$. □

COROLLARY 2.11. *The determinant of a permutation matrix is* ± 1.

PROOF. By part (b') of Lemma 2.5. □

DEFINITION 2.11. *The determinant of a permutation matrix P is called the* ***sign of a permutation****:*

$$sign(p) = \det P = \pm 1.$$

A permutation is called **even** *or* **odd** *according to whether its sign is* $+1$ *or* -1.

For instance, any transposition, i.e., an elementary matrix of Type III (interchange of two rows), has *sign* -1.

We are now ready to derive the *complete expansion* of the determinant. Let $A = (a_{ij})$ be an $n \times n$ matrix and we expand it through its first row, and we do the same thing inductively for the intermediate submatrices:

$$\begin{aligned} \det(A) &= \det \begin{bmatrix} a_{11}, 0, ..., 0 \\ R_2 \\ \vdots \\ R_n \end{bmatrix} \\ &\quad + \det \begin{bmatrix} 0, a_{12}, 0, ..., 0 \\ R_2 \\ \vdots \\ R_n \end{bmatrix} + \cdots + \det \begin{bmatrix} 0, ..., 0, a_{1n} \\ R_2 \\ \vdots \\ R_n \end{bmatrix} \\ &= \sum \det(M), \end{aligned}$$

where the matrix

$$M = \begin{bmatrix} & & a_{1?} & & \\ & a_{2?} & & & \\ & & & & \\ & & & & \\ & & & a_{n?} & \end{bmatrix}$$

having no column zero must have one entry a_{ij} left in each row and column. We note the there are exactly $n!$ such matrices M among n^n of them.

EXAMPLE 2.21. *For a* 2×2 *matrix* A, *we have*

$$\begin{aligned} \det\begin{bmatrix} a & b \\ c & d \end{bmatrix} &= \det\begin{bmatrix} a & 0 \\ c & d \end{bmatrix} + \det\begin{bmatrix} 0 & b \\ c & d \end{bmatrix} \\ &= \det\begin{bmatrix} a & 0 \\ c & 0 \end{bmatrix} + \det\begin{bmatrix} a & 0 \\ 0 & d \end{bmatrix} \\ &\quad + \det\begin{bmatrix} 0 & b \\ c & 0 \end{bmatrix} + \det\begin{bmatrix} 0 & b \\ 0 & d \end{bmatrix} \\ &= \det\begin{bmatrix} a & 0 \\ 0 & d \end{bmatrix} + \det\begin{bmatrix} 0 & b \\ c & 0 \end{bmatrix}. \end{aligned}$$

In particular we note that these matrices M are like permutation matrices P, except that the 1's in P are replaced with the entries of A:

$$P = \sum_j e_{p(j)j}, \qquad M = \sum_j a_{p(j)j} e_{p(j)j} \tag{2.25}$$

Then by linearity of the determinant, we see that

$$\begin{aligned} \det(M) &= a_{p(1)1} \cdots a_{p(n)n} \det P \\ &= (sign\ p) \cdot (a_{p(1)1} \cdots a_{p(n)n}). \end{aligned}$$

There is one such term for each permutation p. Therefore, we finally have

$$\begin{aligned} \det(A) &= \sum \det(M) \\ &= \sum_{perm\ p} (sign\ p) \cdot a_{p(1)1} \cdots a_{p(n)n} \\ &= \sum_{perm\ p} (sign\ p) \cdot a_{1p(1)} \cdots a_{np(n)}, \end{aligned} \tag{2.26}$$

since $\det(A) = \det(A^t)$, where the sum is taken over all permutations of the set $\{1, 2, ..., n\}$. In this way, we have established the complete expansion of the determinant:

$$\det(A) = \sum_{perm\ p} (sign\ p) \cdot a_{1p(1)} \cdots a_{np(n)} \ . \tag{2.27}$$

□

EXAMPLE 2.22.

$$\det \begin{bmatrix} a_{11} & a_{12} & a_{13} \\ a_{21} & a_{22} & a_{23} \\ a_{31} & a_{32} & a_{33} \end{bmatrix} = \begin{array}{c} a_{11}a_{22}a_{33} + a_{12}a_{23}a_{31} + a_{13}a_{21}a_{32} \\ -a_{11}a_{23}a_{32} - a_{12}a_{21}a_{33} - a_{13}a_{22}a_{31}. \end{array}$$

Exercises for Section 2.5

EXERCISE 2.37. *Let p be a permutation on $\mathbf{n} = \{1, 2, 3, 4\}$ defined by sending* $1 \mapsto 3 \mapsto 4 \mapsto 2 \mapsto 1$.

(1) *What is its corresponding permutation matrix?*
(2) *Can you write it as a product of transpositions and evaluate the corresponding matrix product?*
(3) *What is the sign of p?*

EXERCISE 2.38. *Prove that every permutation matrix is a product of transpositions.*

EXERCISE 2.39. *Prove that sign p = sign p^{-1} for any permutation p.*

EXERCISE 2.40. *Prove that a matrix with a single* 1 *in each row and a single* 1 *in each column, but the other entries being zero, is a permutation matrix. (Hint, You start with the identity matrix.)*

EXERCISE 2.41. *Prove that the complete expansion formula in 2.27 defines the determinant function by directly verifying those three rules in Definition 2.8.*

EXERCISE 2.42. *Compute the determinant of the following matrix using the complete expansion formula in 2.27:*

$$\det \begin{bmatrix} 1 & 2 & 5 & 6 \\ 3 & 1 & 7 & 7 \\ 0 & 0 & 2 & 3 \\ 4 & 2 & 1 & 5 \end{bmatrix}.$$

EXERCISE 2.43. *Compute the determinant of the above matrix using those rules in Lemma 2.5.*

6. Cramer's Rule

The name *Cramer's Rule* is applied to a group of formulas giving solutions of systems of linear equations in terms of determinants; thus they are closely related to the minor-expansion of it through the jth column and through the ith row, recursively:

Expansion by minors on the jth column:

$$\begin{aligned} (2.28)\ \det(A) &= (-1)^{1+j}a_{1j}\det(A_{1j}) + (-1)^{2+j}a_{2j}\det(A_{2j}) + \ldots \\ &\quad +(-1)^{n+j}a_{nj}\det(A_{nj}). \end{aligned}$$

Expansion by minors on the ith row:

$$\begin{aligned} (2.29)\ \det(A) &= a_{i1}(-1)^{i+1}\det(A_{i1}) + a_{i2}(-1)^{i+2}\det(A_{i2}) + \ldots \\ &\quad +a_{in}(-1)^{i+n}\det(A_{in}). \end{aligned}$$

EXERCISE 2.44. *Prove the formula in Eq. (2.28).*

Once the first one is proved, the second one can be derived from it by transposing the matrix and applying Proposition 2.6.

We now introduce one more

DEFINITION 2.12. *Let A be an $n \times n$ matrix. The **adjoint** of A, adj A, is the $n \times n$ matrix whose (i, j) entry $(adj)_{ij}$ is $(-1)^{j+i}\det(A_{ji}) = \alpha_{ji}$, where A_{ij} is the matrix obtained by crossing out the ith row and j^{th} column of A. That is,*

$$adj\ A = (\alpha_{ij})^t,$$

where $\alpha_{ij} = (-1)^{i+j}\det(A_{ij})$.

EXAMPLE 2.23. $adj \begin{bmatrix} a & b \\ c & d \end{bmatrix} = \begin{bmatrix} d & -b \\ -c & a \end{bmatrix}$.

THEOREM 2.3.

$$(adj\ A)\cdot A = \det(A)I, \quad \text{and} \quad A\cdot(adj\ A) = \det(A)I. \tag{2.30}$$

PROOF. We first note that the (i,j) entry of the matrix $(adj\ A)\cdot A$ is

$$(adj)_{i1}a_{1j} + \cdots + (adj)_{in}a_{nj} = \alpha_{1i}a_{1j} + \cdots + \alpha_{ni}a_{nj}. \tag{2.31}$$

If $i = j$, i.e., for the diagonal entries, this is formula (2.28) for $\det(A)$, thus we are done. Otherwise, i.e., if $i \neq j$, we are left to show that the (i,j) entries are all zero. For this, we consider the matrix B obtained by replacing the ith column with the jth column in the matrix A. Then Eq. (2.31) is just expansion by minors for B on its ith column; hence $\det(B) = 0$ since B has the same column twice in i^{th} and j^{th} columns. This proves the first equality. Similarly, the second equality is proved. □

COROLLARY 2.12. *If A is invertible, then*

$$A^{-1} = \frac{1}{\det(A)}(adj\ A).$$

EXAMPLE 2.24. *If* $\det(A) \neq 0$, *then* $\begin{bmatrix} a & b \\ c & d \end{bmatrix}^{-1} = \frac{1}{\det(A)}\begin{bmatrix} d & -b \\ -c & a \end{bmatrix}$.

Formula (2.30) can be used to write the solution of a system of linear equations $AX = B$, where A is an $n \times n$ matrix, in a compact form, provided that $\det(A) \neq 0$.

Multiplying both sides by A^{-1}, we get

$$\begin{bmatrix} x_1 \\ x_2 \\ \vdots \\ x_n \end{bmatrix} = A^{-1}B = \frac{1}{\det(A)}(adj\ A)B,$$

and hence we get the formula for each x_j:

$$x_j = \frac{1}{\det(A)}(b_1\alpha_{1j} + \cdots + b_n\alpha_{nj}), \tag{2.32}$$

where $\alpha_{ij} = (-1)^{i+j}\det(A_{ij})$, as above. If we define a new matrix M_j for each j to be the matrix obtained from A by replacing the jth column of A with the column vector B, then

$$\det(M_j) = b_1\alpha_{1j} + \cdots + b_n\alpha_{nj},$$

and hence we get the well-known formula for x_j:

$$x_j = \frac{\det(M_j)}{\det(A)}. \tag{2.33}$$

REMARK 2.10. *For some reason it is popular to write the solution of the system of linear equations $AX = B$ in this form, and call it* Cramer's Rule. *But, this formula is computationally not very good. On the other hand, like the complete expansion of the determinant (2.27), formulas derived in this section have theoretical significance, because the answers A^{-1} and X are exhibited explicitly as quotients of polynomials in the variables $\{a_{ij}, b_i\}$, with integer coefficients. Thus, if, for instance, they are all continuous functions of a single variable t, then so are the solutions x_i.*

EXAMPLE 2.25. *(MAPLE) The following lines compute the adjoint of a square matrix:*

```
interface(echo=2);
with(linalg):
a:=matrix([[1,0,0,0,1,0,0,0],[5,2,0,0,0,1,0,0], \
[8,6,3,0,0,0,1,0],[0,9,7,4,0,0,0,1]]);
aa:=submatrix(a,1..4,1..4);
id:=matrix([[1,0,0,0],[0,1,0,0],[0,0,1,0],[0,0,0,1]]);
b:=adj(aa);
# We can verify this result using the formula in Eq. (2.30).
evalm(b&*aa-det(aa)*id);
```

Exercises for Section 2.6

EXERCISE 2.45. *What is* A^{-1} *of the matrix*

$$A = \begin{bmatrix} 1 & 2 & 3 \\ 4 & 5 & 6 \\ 7 & 8 & 9 \end{bmatrix}?$$

Is Cramer's Rule an easy way of computing it?

EXERCISE 2.46. *Let* A *be an* $n \times n$ *matrix with integer entries* a_{ij}. *Prove that* A^{-1} *has integer entries if and only if* $\det(A) = \pm 1$.

EXERCISE 2.47. *Consider a system of* n *linear equations in* n *unknowns:* $AX = B$, *where* A *and* B *have integer entries. Prove or disprove the followings:*

(1) The system has a rational solution if $\det(A) \neq 0$.
(2) If the system has rational solution, then it also has an integer solution.

CHAPTER 3

Real Vector Spaces

1. Subspaces

A vector space V over the field $\mathbb{R}$ of real numbers is called a *real vector space.*

EXAMPLE 3.1.

(a): *The set $\mathbb{R}^n$ of n-tuples with real entries is a real vector space under vector addition and scalar multiplication:*

$$\begin{bmatrix} a_1 \\ \vdots \\ a_n \end{bmatrix} + \begin{bmatrix} b_1 \\ \vdots \\ b_n \end{bmatrix} = \begin{bmatrix} a_1 + b_1 \\ \vdots \\ a_n + b_n \end{bmatrix},$$

$$c \begin{bmatrix} a_1 \\ \vdots \\ a_n \end{bmatrix} = \begin{bmatrix} ca_1 \\ \vdots \\ ca_n \end{bmatrix}.$$

(b): *The field $\mathbb{C}$ of complex numbers is a real vector space.*

(c): *The set of real polynomials $p(x) = a_n x^n + \cdots + a_0$ is a real vector space.*

(d): *The set of continuous real-valued functions on the interval $[0, 1]$ is a real vector space under the usual operations on such functions:*

$$\begin{aligned} (f + g)(x) &= f(x) + g(x) \\ (cf)(x) &= cf(x). \end{aligned}$$

REMARK 3.1. *Multiplication of two vectors is not a part of the structure, though various products, such as the* cross product *of vectors in $\mathbb{R}^3$, can be defined. These products are not completely intrinsic; they depend* *on choosing coordinates. They are considered to be additional structure on the vector space.*

DEFINITION 3.1. *A* non-empty *subset W of a vector space V over a field F is called a* **subspace** *of V if*

(1) *For each elements w_1 and w_2 of W, their sum $w_1 + w_2 \in W$.*
(2) *For each scalar α and each $w \in W$, their scalar multiplication $\alpha w \in W$.*

A subspace W of V is called *proper* if it is neither the whole space V nor the zero space $\{0\}$.

EXAMPLE 3.2. *The subspaces of $\mathbb{R}^3$ are of four types:*

(1) *the zero vector.*
(2) *the vectors lying on a line through the origin $(0, 0, 0)$.*
(3) *the vectors lying on a plane through the origin.*
(4) *the whole space $\mathbb{R}^3$.*

Such a classification of subspaces of $\mathbb{R}^3$ is based on the concept of dimension, which we will learn later in this chapter.

EXAMPLE 3.3. *The set of solutions of a system of homogeneous linear equations is a subspace of $\mathbb{R}^n$ for some n. This is because, for a matrix A of size $m \times n$, we easily see:*

(1) *$AX = 0$ and $AY = 0$ implies $A(X + Y) = 0$, that is, if X and Y are solutions, then so is their sum $X + Y$.*
(2) *$AX = 0$ implies $AcX = 0$, that is, if X is a solution, then so is cX.*

Here is a little bit advanced

EXAMPLE 3.4. *Let W be the set of solutions of a single equation:*

$$2x - y - z = 0.$$

Thus, W is a collection of those vectors $w = (x, y, z)$ in $\mathbb{R}^3$ that are perpendicular to the column vector $v = (2, -1, -1)^t$, i.e., satisfying the *relationship* $wv = 0$ *defined in 2.6 (Also see Lemma 4.7.) It is going to be a plane through the origin since it is a solution set of a* single *linear equation with* three *unknowns. Thus we know that there are infinitely many such vectors. In fact, by putting* $z = t$ *(a parameter),* $y = u$ *(another parameter), we solve it for* x*:* $x = 1/2u + 1/2t$ *and hence its solution set consists of*

$$\begin{bmatrix} x \\ y \\ z \end{bmatrix} = u \begin{bmatrix} 1/2 \\ 1 \\ 0 \end{bmatrix} + t \begin{bmatrix} 1/2 \\ 0 \\ 1 \end{bmatrix},$$

where u *and* t *are arbitrary real numbers (i.e., parameters),* linear combinations *of two vectors* $w_1 = (1/2, 1, 0)^t$ *and* $w_2 = (1/2, 0, 1)^t$*. Since* w_1 *is not a non-zero scalar multiple of* w_2*, there is no way of expressing* w_1 *in terms of* w_2*, and hence they are called* linearly independent. *Another point is that once we know these two vectors, we can express all other solutions in terms of these two, and when this happens, we say that they* span (or generate) *the* solution space *W of the equation. Note that we already know that* W *is a subspace of* $\mathbb{R}^3$.

LEMMA 3.1. *Let* V *be a real vector space. Then for each* $v \in V$ *and for each* $c \in \mathbb{R}$,

(a): $0_{\mathbb{R}} v = 0_V$;
(b): $c 0_V = 0_V$;
(c): $(-1)v = -v$,

where $0_{\mathbb{R}}$ *is the additive identity of* $\mathbb{R}$ *and* 0_V *is the additive identity of* V.

PROOF. To see (a), we use the distributive law:

$$0v + 0v = (0 + 0)v = 0v = 0v + 0.$$

Cancelling $0v$ from both sides, we obtain $0v = 0$.

Similarly, $c0 + c0 = c(0 + 0) = c0$. Thus, $c0 = 0$. Finally,

$$v + (-1)v = (1 + (-1))v = 0v = 0.$$

Thus, $(-1)v$ is an additive inverse of v. Then by uniqueness of such an inverse, it must be the same as the additive inverse $-v$ of v. □

When we want to compare two sets we use maps. In particular, a *bijective map*, i.e., a one-to-one and onto map, between two vectors spaces with a special property is called an isomorphism:

DEFINITION 3.2. *An* ***isomorphism*** φ *from a vector space* V *to a vector space* V', *both over the same field* F, *is a bijective map* $\varphi : V \longrightarrow V'$ *compatible with the laws of composition, that is, a bijective map satisfying: for each* $v_1, v_2 \in V$ *and for each* $c \in F$,

(1) $\varphi(v_1 + v_2) = \varphi(v_1) + \varphi(v_2)$;
(2) $\varphi(cv) = c\varphi(v)$.

EXAMPLE 3.5. *The map* $\varphi : \mathbb{R}^2 \longrightarrow \mathbb{C}$ *sending* $(a, b) \mapsto a + bi$, *where* $i = \sqrt{-1}$, *is an isomorphism of* $\mathbb{R}$*-vector spaces* $\mathbb{R}^2$ *and* $\mathbb{C}$.

LEMMA 3.2. *The intersection of two subspaces* W_1, W_2 *of a vector space* V *is again a subspace of* V. □

Exercises for Section 3.1

EXERCISE 3.1. *Prove that the intersection of two subspaces is a subspace.*

EXERCISE 3.2. *Prove the cancellation law holds in a vector space: If* $cv = cw$ *and* $c \neq 0$ *then* $v = w$.

EXERCISE 3.3. *Is* $-w \in W$ *if* w *is a vector in a subspace* W?

EXERCISE 3.4. *Prove that the set of symmetric matrices of size* $n \times n$ *is a subspace of the vector space consisting of all real* $n \times n$ *matrices.*

EXERCISE 3.5. *Prove that the set of upper triangular matrices of size* $n \times n$ *is a subspace of the vector space consisting of all real* $n \times n$ *matrices.*

2. Bases and Dimension

In this section we introduce a few more terminologies for an abstractly given vector space. The new concepts are *span*, *linear independence*, and *basis*.

REMARK 3.2. *When computations are involved, we need to* order *the elements of a given set of vectors. We have been putting curly brackets around unordered sets, so in order to distinguish ordered from unordered sets, we are going to use rounded brackets to enclose ordered sets. Thus, by definition, the ordered set (a, b) is considered to be different from the ordered set (b, a), whereas the unordered sets $\{a, b\}$ and $\{b, a\}$ are considered to be the same. In particular, repetitions will also be allowed in an ordered set, so (a, a, b) is an ordered set, and it is different from (a, b), in contrast to the convention for unordered sets, where, for instance, $\{a, a, b\}$ and $\{a, b\}$ are exactly the same.*

Let V be a vector space over a field F, and let $(v_1, ..., v_n)$ be an *ordered set* of elements of V.

DEFINITION 3.3. *A* ***linear combination*** *of $\{v_1, ..., v_n\}$ is any vector of the form $w = c_1v_1 + c_2v_2 + \cdots + c_nv_n, \qquad c_i \in F$.*

DEFINITION 3.4. *For any scalar c and a vector v, we define $vc := cv$.*

Thus, $w = v_1c_1 + \cdots + v_nc_n$ is also a linear combination of the vectors $v_1, ..., v_n$. Later we will see the usefulness of this definition when we use matrix multiplication very often.

EXAMPLE 3.6.

(a): *Let $v_1 = (1, 0, 2)^t$ and $v_2 = (3, 3, 0)^t$ be two vectors in $\mathbb{R}^3$. Then the vector $(5, 3, 4)^t = 2v_1 + v_2$ is a linear combination of v_1 and v_2. In general, $c_1v_1 + c_2v_2 = (c_1 + c_2, 3c_2, 2c_1)^t$ is a linear combination of v_1 and v_2.*

(b): *A solution X of a system of linear equations written in the matrix form $AX = B$ exhibits the column vector B as a linear* *combination of the columns of the matrix* A *of size* $m \times n$. *The coefficients are the entries of the vector* X. *That is, writing* $A = (v_1, v_2, ..., v_n)$, $X = (x_1, ..., x_n)^t$ *and* $B = (b_1, ..., b_m)^t$, *we see that*

$$\begin{bmatrix} b_1 \\ \vdots \\ b_m \end{bmatrix} = x_1 v_1 + \cdots + x_n v_n,$$

where each v_i *is the* i^{th} *column of the matrix* A.

(c): *A linear combination of a single vector* v *is just a scalar multiple* cv *of* v.

LEMMA 3.3. *The set* $Span(S)$ *of all vectors which are linear combinations of the vectors* $S = \{v_1, ..., v_n\}$ *of a vector space* V *is a subspace of* V.

PROOF. Obviously, $0 = 0v_1 + ... + 0v_n$, so it is an element of $Span(S)$. If $w = c_1 v_1 + ... + c_n v_n$ and $w' = c_1' v_1 + ... + c_n' v_n$ are two vectors in $Span(S)$, then their sum $w + w' = (c_1 + c_1')v_1 + ... + (c_n + c_n')v_n$ is a linear combination of them; and hence it is an element of $Span(S)$, and if $a \in F$, then $aw = ac_1 v_1 + ... + ac_n v_n \in Span(S)$; hence $Span(S)$ is indeed a subspace of V. □

The space spanned by a set S is denoted by $Span(S)$. We note that $Span(S)$ is clearly the smallest subspace of V containing S. We also call $Span(S)$ the space **generated by** S. We also note that the order among the vectors $v_1, ..., v_n$ are not important here: $Span(S)$ is the same as the space of any reordering of elements of S.

PROPOSITION 3.1. *Let* S *be a set, finite or not, of vectors of* V, *and let* W *be a subspace of* V. *If* $S \subseteq W$, *then* $Span(S) \subseteq W$.

PROOF. This is because W is closed under vector addition and scalar multiplication. □

The following are perhaps one of the most important concepts in Linear Algebra:

DEFINITION 3.5.

(a): *A* ***linear relation*** *among vectors* $v_1, ..., v_n$ *is any relation of the form:*

$$c_1v_1 + c_2v_2 + ... + c_nv_n = 0,$$

where the coefficients c_i *are elements of* F.

(b): *The vectors* $v_1, ..., v_n$ *are (or equivalently, a set consisting of these vectors is)* ***linearly independent*** *if* $c_1v_1+c_2v_2+...+c_nv_n = 0$ *implies each coefficients* $c_i = 0$.

(c): *The vectors* $v_1, ..., v_n$ *(or a subset consisting of them) are* ***linearly dependent*** *if they are not linearly independent.*

(d): *A set* S*, finite or not, is linearly independent if it does not have any linearly dependent subsets.*

(e): *A set* S*, finite or not, is linearly dependent if it is not linearly independent.*

The following example will show how we can check whether given set of vectors are linearly independent or not, provided that they are given explicitly.

EXAMPLE 3.7. *Let* $v_1 = (1,0,1)^t, v_2 = (1,1,0)^t$, *and* $v_3 = (1,1,1)^t$. *We want to check whether they are linearly independent or not. For this, we consider a matrix*

$$A^t = [v_1, v_2, v_3]^t = \begin{bmatrix} 1 & 0 & 1 \\ 1 & 1 & 0 \\ 1 & 1 & 1 \end{bmatrix}.$$

We apply elementary row operations to A^t *to obtain the row reduced echelon form* B *of it. If the bottom row of* B *is zero, this means that one of the three vectors is a linear combination of the other two vectors with non-zero coefficients, and hence the original vectors are linearly dependent. Otherwise, the original vectors are linearly independent. In the case of this example, we see that* $B = I$*, and hence they are linearly independent.*

EXAMPLE 3.8. *We work with the same vectors in the above Example. This time, we consider a system of homogeneous linear equations:*

$$AX = 0.$$

This expression is exactly the same as

$$\begin{bmatrix} v_1 & v_2 & v_3 \end{bmatrix} \begin{bmatrix} x_1 \\ x_2 \\ x_3 \end{bmatrix} = v_1x_1 + v_2x_2 + v_3x_3 = 0,$$

it is simply another way of expressing a linear relation among the three vectors $v_1, v_2,$ *and* v_3. *Therefore, if we can show that the zero-vector* $(x_1, x_2, x_3)^t$ *is the only solution of the system, then the original vectors are linearly independent. Otherwise, they are linearly dependent. In the case of this example,* $\det(A) \neq 0$, *i.e.,* A *is invertible, and hence the vectors are linearly independent.*

On the other hand, if we add an arbitrary fourth vector v_4 *to this set, then they are going to be linearly dependent because every system of three homogeneous linear equations in four unknowns has a non-trivial solution.*

We have an easy proposition whose proof is left to the readers as an exercise.

PROPOSITION 3.2.

- **(a):** *Any reordering of a linearly independent set is linearly independent.*
- **(b):** *If* $v \in V$ *is a non-zero vector, then the set* $\{v\}$ *is linearly independent.*
- **(c):** *A set* $\{v_1, v_2\}$ *of two vectors is linearly dependent if and only if either one of them is a zero vector or one of them is a scalar multiple of the other.*

DEFINITION 3.6. *A set of vectors* $v_1, ..., v_n$ *of a vector space* V *which is linearly independent and which also spans* V *is called a* ***basis*** *of* V.

EXAMPLE 3.9. *The two vectors* w_1 *and* w_2 *in Example 3.4 are a basis of the solution space they span.*

PROPOSITION 3.3. *Let* $B = \{v_1, ..., v_n\}$ *be a set of vectors in a vector space* V. *Then it is a basis of* V *if and only if every vector* $w \in V$ *can be written in a unique way as a linear combination of them.*

PROOF. The $\Longleftarrow$ direction is obvious. For the other direction, suppose there is a vector w in V having two expressions:

$$w = c_1v_1 + ... + c_nv_n = c_1'v_1 + ... + c_n'v_n.$$

Then since they are linearly independent,

$$0 = (c_1 - c_1')v_1 + ... + (c_n - c_n')v_n$$

means that each $c_i = c_i'$. □

EXAMPLE 3.10. *The vectors* $e_1, ..., e_n$ *in* $\mathbb{R}^n$ *is a basis of* $\mathbb{R}^n$, *called the* standard basis, *because every vector* $w = (x_1, ..., x_n)^t$ *in* $\mathbb{R}^n$ *can be written uniquely as*

$$(x_1, ..., x_n)^t = x_1e_1 + ... + x_ne_n,$$

as a linear combination of the e_i*'s.*

PROPOSITION 3.4. *Let* L *be a linearly independent subset of* V, *and let* $v \in V$ *be any non-zero vector. Then the set* $L' = L \cup \{v\}$ *is linearly independent if and only if* v *is not in* $Span(L)$.

PROOF. If $v \in Span(L)$ then L' is obviously linearly dependent. Conversely, suppose that L' is linearly dependent. Then, since L is linearly independent by hypothesis, there exist, by the part (e) of 3.5, vectors $v_1, ..., v_r$ in L' such that

$$c_1v_1 + ... + c_rv_r + cv = 0,$$

with some non-zero coefficients. If $c = 0$, then $c_1v_1 + ... + c_rv_r = 0$; hence each $c_i = 0$ for they are linearly independent by hypothesis; consequently L' is linearly independent, which is contrary to our assumption. Thus, $c \neq 0$; hence v is a linear combination of the v_i's:

$$v = \frac{-c_1}{c}v_1 + ... + \frac{-c_r}{c}v_r.$$

Thus, $v \in Span(L)$. □

PROPOSITION 3.5. *Let S be a subset of V, and let $v \in V$ be any vector. Then $Span(S) = Span(S \cup \{v\})$ if and only if $v \in Span(S)$.*

PROOF. Since $v \in Span(S \cup \{v\})$, if $v \notin Span(S)$ then $Span(S \cup \{v\}) \neq Span(S)$. Conversely, if $v \in Span(S)$, then $S \cup \{v\} \subseteq Span(S)$; hence $Span(S \cup \{v\}) \subseteq Span(S)$ by Proposition 3.1; hence $Span(S) = Span(S \cup \{v\})$. □

DEFINITION 3.7. *A vector space V is called **finite-dimensional** if there is a finite subset of V which spans V.*

From now on, we assume, unless otherwise stated, that our given vector space is finite-dimensional.

PROPOSITION 3.6. *Any finite set S which spans V contains a basis. In particular, any finite-dimensional vector space has a basis.*

PROOF. Since the zero vector 0 is nothing to do with a basis of V, by deleting 0 from the set S if S has it, we may assume that $0 \notin S$. If $S = \{v_1, ..., v_n\}$ is linearly independent, then we are done. Otherwise, there is a linear relation

$$c_1v_1 + ... + c_nv_n = 0,$$

with some $c_i \neq 0$. We may assume that $c_n \neq 0$; then we may write v_n as a linear combination of other vectors, and hence $v_n \in Span(S')$ where $S' = S - \{v_n\}$. Thus, $V = Span(S) = Span(S')$, by Proposition 3.5. If S' is linearly independent, then we are done. Otherwise, we repeat the above process again. Continuing this way we eventually obtain a family which is linearly independent but still spans V. □

REMARK 3.3. *In order to avoid possible misleading with the above proof when $V = \{0\}$, we adapt the followings:*

a: *The empty set is linearly independent.*

b: *The span of the empty set is the zero subspace.*

PROPOSITION 3.7. *Let V be a finite-dimensional vector space. Any linearly independent finite subset L of V can be extended to a basis of V.*

PROOF. Let S be a finite set which spans V. If $S \subseteq Span(L)$, then $V = Span(S) \subseteq Span(L) \subseteq V$; hence L is a basis. Otherwise, choose a non-zero vector $v \in S - Span(L)$. Then by Proposition 3.4, $L \cup \{v\}$ is linearly independent. We continue this process until we get a basis of V after finitely many steps. □

PROPOSITION 3.8. *If a finite subset S of V spans V and a finite subset L of V is linearly independent, then $|S| \geq |L|$.*

PROOF. Write $S = \{v_1, ..., v_m\}$ and $L = \{w_1, ..., w_n\}$. Then since S generates V, we write each w_j as a linear combination of S:

$$w_j = a_{1j}v_1 + ... + a_{mj}v_m = \sum_{i=1}^{m} a_{ij}v_i.$$

Consider $u = c_1w_1 + ... + c_nw_n$. By replacing each w_j in this expression, we get

$$u = \sum_{i,j} c_j a_{ij} v_i = (\sum_j a_{1j}c_j)v_1 + ... + (\sum_j a_{mj}c_j)v_m, \tag{3.1}$$

where the coefficient of v_i is $\sum_j a_{ij}c_j$. Since L is linearly independent, we know that $u = 0$ implies each $c_j = 0, j = 1, 2, ..., n$.

(By Contradiction) Now, let us examine Eq. (3.1) carefully in order to see what happens if $m < n$: We note that we have a system of m homogeneous linear equations with n unknowns:

$$\sum_{j=1}^{n} a_{ij}x_j = 0, \qquad i = 1, 2, ..., m,$$

which we obtain from Eq. (3.1). If $m < n$, then this system would have a non-trivial solution $(c_1, ..., c_n)^t$, by Corollary 2.1 of Chapter 2; consequently, with these some non-zero c_j's, we would have $u = c_1w_1 + ... + c_nw_n = 0$ via Eq. (3.1). But, this is impossible because L is linearly independent by hypothesis. Therefore, we must always have $m \geq n$. □

PROPOSITION 3.9. *Two bases B and B' of the vector space V have the same number of elements.*

PROOF. By Proposition 3.8, $|B| \geq |B'|$, and $|B'| \geq |B|$. $\square$

REMARK 3.4. *This proposition can be proved even when V is not a finite-dimensional vector space. The only problem in this case is that we then have to use the axiom of choice in proving it generally.*

EXAMPLE 3.11. *Let $V = \mathbb{R}[X]$ be the set of all polynomials with coefficients in $\mathbb{R}$. It is easy to check that it is a vector space over $\mathbb{R}$ and the set $\{X^i : i = 0, 1, 2, , , \}$ is a basis of V.*

Because of these results we see that the number of elements of a basis of a vector space is *intrinsic.*

DEFINITION 3.8. *The **dimension** of a finite-dimensional vector space V is the number of vectors in a basis. The dimension will be denoted by* $\dim V$.

PROPOSITION 3.10.

- **a:** If S spans V, then $|S| \geq \dim V$, and equality holds only if S is a basis. That is, if S is not a basis, then equality does not hold.
- **b:** If L is linearly independent, then $|L| \leq \dim V$, and equality holds only if L is a basis.

PROOF. They follow from Proposition 3.6 and Proposition 3.7. $\square$

REMARK 3.5. *It is obvious that if L is a linearly independent subset of a subspace W of a vector space V, then L is also a linearly independent subset of V when viewed as a subset of V.*

PROPOSITION 3.11. *If $W \subseteq V$ is a subspace of a finite-dimensional vector space V, then W is also finite-dimensional, and* $\dim W \leq \dim V$. *Moreover,* $\dim W = \dim V$ *only if $W = V$.*

PROOF. We start with the empty set and add elements of W using Proposition 3.4, in order to end up with a basis of W. By the above

Remark and Proposition 3.8, we see that any linearly independent subset L of W satisfies: $|L| \leq n = \dim V$. Thus the process of adding vectors of W to L must stop after at most n steps. When it is impossible to apply Proposition 3.4 any more, L is a basis of W; and hence W is finite-dimensional and $\dim W \leq \dim V$.

Now if W is a proper subset of V, then a basis of W cannot be a basis of V, but it can be extended to a basis of V by Proposition 3.7. Therefore, $\dim W < \dim V$, in this case. □

PROPOSITION 3.12. *Let V be a finite-dimensional vector space, and let S be any set (finite or infinite) which spans V. Then S contains a finite subset spanning V.*

PROOF. By hypothesis, there exists a finite set, say $\{w_1, ..., w_m\}$, which spans V. Since $Span(S) = V$, each w_i is a linear combination of finitely many elements of S. Thus when we express the vectors $w_1, ..., w_m$ in terms of the set S, we only need finitely many elements of S. If S' is the collection of them, then $\{w_1, ..., w_m\} \subset Span(S')$; hence S' spans V. □

PROPOSITION 3.13. *Let V be a finite-dimensional vector space. Then*

- **a:** *Every set S which spans V contains a finite basis.*
- **b:** *Every linearly independent set L is finite and therefore extends to a finite basis.*
- **c:** *Every basis is finite.*

PROOF. By Proposition 3.12 and Proposition 3.6, (a) holds. By part (a) and Proposition 3.9, (c) holds. Then finally by part (c), part (b) of Proposition 3.10, and Proposition 3.7, (b) holds. □

EXAMPLE 3.12. *(MAPLE) The following lines compute a basis of the subspace generated by a given set of vectors v1, v2, and v3:*

———————

interface(echo=2);

```
with(linalg):
v1:=vector([-1,2,3,0]);
v2:=vector([0,1,2,1]);
v3:=vector([2,-1,-1,1]);
basis( {v1, v2, v3} );
basis ({v2,v1,v3});
```

———————

EXAMPLE 3.13. *(MAPLE) The following lines compute a basis of the subspace generated by a given set of vectors v1, v2, and v3:*

———————

```
interface(echo=2);
with(linalg):
basis( {vector([1,1,1]), vector([2,2,2]), vector([1,-1,1]),
vector([2,-2,2]), vector([1,0,1]), vector([0,1,1])} );
```

———————

Exercises for Section 3.2

EXERCISE 3.6. *Prove that a subset of a linearly independent set is also linearly independent.*

EXERCISE 3.7. *Prove that any reordering of a basis is also a basis.*

EXERCISE 3.8. *Find a basis for the subspace of $\mathbb{R}^4$ spanned by the vectors*

$$(1,2,1,1), (3,2,-1,3), (1,0,3,4), (2,5,3,2).$$

EXERCISE 3.9. *Let $V \subset \mathbb{R}^4$ be the solution space of the system of linear equations $AX = 0$, where*

$$A = \begin{bmatrix} 2 & 1 & 3 & 4 \\ 1 & 0 & 2 & 1 \end{bmatrix}.$$

(1) *Find a basis for V.*
(2) *Write down each solution in terms of the basis elements you computed in part (1).*

EXERCISE 3.10. *Find a basis for the space of symmetric $n \times n$ matrices.*

EXERCISE 3.11. *Prove that a square matrix A is invertible if and only if its columns are linearly independent.*

EXERCISE 3.12. *Let A be an $m \times n$ matrix, and let A' be the result of a sequence of elementary row operations on A. Prove that the rows of A spans the same subspace as the rows of A'.*

EXERCISE 3.13. *Let V be a vector space of dimension n over a field F. Prove that V contains a subspace of dimension r for any $0 \leq r \leq n$.*

EXERCISE 3.14. *Let V be a complex vector space of dimension n. Prove that V has dimension $2n$ as real vector space.*

EXERCISE 3.15. *Let $v_1 = (a_1, a_2, a_3)$, $v_2 = (b_1, b_2, b_3)$, $v_3 = (c_1, c_2, c_3)$ be three vectors in $\mathbb{R}^3$. Prove that the following statements are equivalent to each other:*

(1) $\det \begin{bmatrix} a_1 & a_2 & a_3 \\ b_1 & b_2 & b_3 \\ c_1 & c_2 & c_3 \end{bmatrix} \neq 0.$

(2) $v_1, v_2,$ *and* v_3 *are linearly independent.*

(3) $\{v_1, v_2, v_3\}$ *is a basis for* $\mathbb{R}^3$.

3. Computation with Bases

Once we are given bases of vector spaces, we can perform useful computations with respects to these bases, which allow us to understand vector spaces more easily. In this section we will deal with two topics: how to express a vector in terms of a given basis, and how to relate two different bases of the same vector space. In particular, our basis is going to be ordered.

Suppose we are given a basis $\mathbf{B} = (v_1, ..., v_n)$ of a vector space V. Then by Proposition 3.3 we know that every vector $v \in V$ can be written down as a linear combination

$$v = x_1 v_1 + ... + x_n v_n, \qquad x_i \in F, \tag{3.2}$$

in exactly one way.

DEFINITION 3.9. *The scalars x_i are called the* ***coordinates*** *of v, with respect to the given basis, and the column vector*

$$X = \begin{bmatrix} x_1 \\ \vdots \\ x_n \end{bmatrix} \in F^n \tag{3.3}$$

is called the ***coordinate vector*** *of v, with respect to the given basis.*

PROPOSITION 3.14. *There exists an isomorphism of vector spaces between a vector space V with a basis $\mathbf{B} = (v_1, ..., v_n)$ and F^n.*

PROOF. Let $v \in V$ be a vector in V. Then by Proposition 3.3 there exist a unique list of scalars $(x_1, ..., x_n)$ such that Eq. (3.2) holds. This establishes a map ϕ from V to F^n: $v \longmapsto X = (x_1, ..., x_n)^t$. Then it can be easily checked that this map is

(1) linear (recall Definition 1.12 in Chapter 1), and
(2) one-to-one, and onto.

□

We can use this isomorphism to introduce *coordinates* into the vector space V. The coordinate vector of a vector v is by definition $X = \phi(v)$.

COROLLARY 3.1. *Every vector space V of dimension n is isomorphic to the space F^n of column vectors.* □

REMARK 3.6. *We note that F^n is not isomorphic to F^m unless $n = m$, because F^n has a basis of n elements, and the number of elements in a basis dependents only on the vector space, not on the choice of a basis. Thus, the finite dimensional vector spaces over a field F are completely classified by Corollary 3.1: Every V is isomorphic to F^n, for some uniquely determined positive integer n. It follows that we will know all about a arbitrary vector space if we study the basic examples of column vectors. This reduces any problem on vector spaces to the familiar algebra of column vectors, once a basis is given to us.*

Thus, let's start with the space of column vectors F^n. Let $\mathbf{B} = (v_1, ..., v_n)$ be a basis of F^n. Since there will be no confusion, we will also write as $\mathbf{B}$ an $n \times n$ matrix whose column vectors are the $v_1, ..., v_n$, that is,

$$\mathbf{B} = [v_1, ..., v_n] \tag{3.4}$$

Thus if $\mathbf{E}$ is the standard basis of F^n then its matrix $\mathbf{E}$ is just the identity matrix.

In order to use the usual matrix multiplication freely, we introduce the following

DEFINITION 3.10. *A scalar multiplication $x_1 v$ is also written as $v x_1$. That is, we define*

$$v x_1 := x_1 v.$$

Then a linear combination $x_1 v_1 + ... + x_n v_n$ of a basis $\mathbf{B} = (v_1, ..., v_n)$ can be expressed as a matrix multiplication very efficiently as follows:

$$\begin{aligned} x_1 v_1 + ... + x_n v_n &= v_1 x_1 + ... + v_n x_n \\ &= [v_1, ..., v_n] \begin{bmatrix} x_1 \\ \vdots \\ x_n \end{bmatrix} \\ &= \mathbf{B}X, \end{aligned} \tag{3.5}$$

where $X = (x_1, ..., x_n)^t$ is the column vector in F^n. The next Proposition says that a matrix of a basis is always invertible:

PROPOSITION 3.15. *Let A be an $n \times n$ matrix with entries in a field F. The columns of A form a basis of F^n if and only if A is invertible.*

PROOF. We denote the ith column of A by v_i. Then for any column vector $X = (x_1, ..., x_n)^t$, the matrix product $AX = v_1 x_1 + ... + v_n x_n$ is a linear combination of the set $\mathbf{B} = (v_1, ..., v_n)$. Thus, this set is linearly independent if and only if $AX = 0$ has the only trivial solution $X = 0$ if and only if A is invertible, where the last equivalence holds by Proposition 2.4 in Chapter 2. Moreover, if the n vectors $\mathbf{B} = (v_1, ..., v_n)$ is linearly independent, then it forms a basis because the dimension of F^n is n, by Proposition 3.11. □

Thus, if a vector $Y = (y_1, ..., y_n)^t$ is given, we can determine its coordinate vector X with respect to the basis $\mathbf{B}$ by solving the equation

$$\mathbf{B}X = Y,$$

$X = \mathbf{B}^{-1}Y$.

PROPOSITION 3.16. *Let* $\mathbf{B} = (v_1, ..., v_n)$ *be a basis of* F^n*, and let* $Y \in F^n$ *be a vector. The coordinate vector of* Y *with respect to the basis* $\mathbf{B}$ *is*

$$X = \mathbf{B}^{-1}Y. \quad \square$$

EXAMPLE 3.14. *Let* $v_1 = (1, 2)^t$ *and* $v_2 = (3, 4)^t$. *Then we easily check the matrix*

$$\mathbf{B} = [v_1, v_2]$$

is invertible and its inverse is:

$$\begin{bmatrix} 1 & 3 \\ 2 & 4 \end{bmatrix}^{-1} = \begin{bmatrix} -2 & \frac{3}{2} \\ 1 & -\frac{1}{2} \end{bmatrix},$$

and hence the coordinate vector X *of* $Y = (1, 3)^t$ *with respect to this new basis* $\mathbf{B} = (v_1, v_2)$ *is:*

$$X = \mathbf{B}^{-1}Y = \begin{bmatrix} -2 & \frac{3}{2} \\ 1 & -\frac{1}{2} \end{bmatrix} \cdot \begin{bmatrix} 1 \\ 3 \end{bmatrix} = \begin{bmatrix} \frac{5}{2} \\ -\frac{1}{2} \end{bmatrix},$$

and this exactly means that $Y = \frac{5}{2}v_1 - \frac{1}{2}v_2$.

We now begin to work on an abstractly given vector space V. As above, we want to use matrix notation to facilitate the manipulation of bases, and the way we have written ordered sets of vectors was chosen with this in mind:

$$(v_1, ..., v_m). \tag{3.6}$$

DEFINITION 3.11. *This array of vectors*

$$(v_1, ..., v_m)$$

is called a ***hypervector****.*

If we consider a hypervector as an $1 \times m$ matrix, we can multiply it to an $m \times n$ matrix $A = (a_{ij})$ of scalars as follows:

$$(v_1, ..., v_m)(a_{ij}) = (w_1, ..., w_n), \tag{3.7}$$

where each

$$w_j = v_1 a_{1j} + ... + v_m a_{mj}. \tag{3.8}$$

Thus each vector w_j is a linear combination of $(v_1, ..., v_m)$. In particular, when the matrix A is a $m \times 1$ column vector $X = (x_1, ..., x_m)^t$,

$$(v_1, ..., v_m)X = v_1 x_1 + ... + v_m x_m = \mathbf{B}X \tag{3.9}$$

is simply a linear combination of the $\mathbf{B} = (v_1, ..., v_m)$ with coefficients $x_1, ..., x_m$.

EXAMPLE 3.15.

$$\begin{aligned} (v_1, v_2)\begin{bmatrix} 3 & 2 & 2 \\ 4 & 1 & 1 \end{bmatrix} &= v_1\begin{bmatrix} 3 & 2 & 2 \end{bmatrix} + v_2\begin{bmatrix} 4 & 1 & 1 \end{bmatrix} \\ &= [v_1 3, v_1 2, v_1 2] + [v_2 4, v_2 1, v_2 1] \\ &= [3v_1 + 4v_2, 2v_1 + v_2, 2v_1 + v_2]. \end{aligned}$$

We now come to a very important computational method: **change of basis**. Identifying a vector space V with the isomorphic vector space F^n is useful only when the given basis is very well suited to the problem at hand. Thus, if a given basis is not good for this purpose, then we will try to change the coordinates by choosing a different basis of V.

Suppose that we are given two bases $\mathbf{B}$ and $\mathbf{B}'$ for the same vector space V, say $\mathbf{B} = (v_1, ..., v_n)$ and $\mathbf{B}' = (v_1', ..., v_n')$. Then we have the identity map

$$Id : (V, \mathbf{B}) \longrightarrow (V, \mathbf{B}'),$$

where, for instance, $(V, \mathbf{B})$ exactly means a vector space V with a basis $\mathbf{B}$. We call $\mathbf{B}$ an old basis and $\mathbf{B}'$ a new basis of V, respectively. On the other hand, there exist isomorphisms of vector spaces: $\phi : (V, \mathbf{B}) \longrightarrow F^n$ which sends each vector $v \in V$ to its coordinate vector $X = (x_1, ..., x_n)^t \in F^n$ with respect to $\mathbf{B}$, and $\phi' : (V, \mathbf{B}') \longrightarrow F^n$ which sends each vector $v \in V$ to its coordinate vector $X' = (x_1^{'}, ..., x_n^{'}) \in F^n$ with respect to $\mathbf{B}'$. If we put these information all together, we have

$$
\begin{array}{ccc}
(V, \mathbf{B}) & \longrightarrow & (V, \mathbf{B}') \\
\phi \downarrow & \circlearrowleft & \downarrow \phi' \\
F^n & \dashrightarrow & F^n
\end{array}
$$

where the two vertical arrows are isomorphisms sending $\phi : v \mapsto X$ and $\phi' : v \mapsto X'$, respectively. Therefore, the composition, $\phi' \circ Id \circ \phi^{-1}$ is a linear map sending $X \in F^n \mapsto X' \in F^n$(the second row), which makes this diagram **commutative**, namely, $(\phi' \circ Id \circ \phi^{-1}) \circ \phi = \phi' \circ Id : (V, \mathbf{B}) \longrightarrow F^n$.

We want to figure out this map $\phi' \circ Id \circ \phi^{-1}$ explicitly:

(1) Under the identity map (the upper horizontal map), the old basis $\mathbf{B}$ is sent to $\mathbf{B}$, so we *write the old basis* $\mathbf{B}$ *as a linear combination of the new basis* $\mathbf{B}'$:

$$\mathbf{B} = \mathbf{B}'P. \tag{3.10}$$

That is, for each basis element $v_j \in \mathbf{B} = (v_1, ..., v_n)$, write it uniquely as a linear combination of basis elements of $\mathbf{B}' = (v_1^{'}, ..., v_n^{'})$:

$$v_j = v_1' P_{1j} + v_2' P_{2j} + \cdots + v_n' P_{nj}.$$

Then by definition, the j-th column of the new matrix P consists of the column vector $(P_{1j}, ..., p_{nj})^t$ for each $j = 1, 2, ..., n$.

(2) Also, under the same identity map, the new basis in the target is sent back to itself, so we write the new basis $\mathbf{B}'$ as a linear

combination of the old basis $\mathbf{B}$:

$$\mathbf{B}' = \mathbf{B}Q. \tag{3.11}$$

Then we easily see that the square matrices P and Q are invertible, because we have

$$\mathbf{B} = \mathbf{B}I = \mathbf{B}'P = \mathbf{B}QP \Longrightarrow QP = I. \tag{3.12}$$

This holds because the basis $\mathbf{B}$ is a linearly independent set.

(3) Under $\phi' \circ Id \circ \phi^{-1}$, any column vector $X \in F^n$ is sent to $PX \in F^n$:

$$X \mapsto v = \mathbf{B}X \mapsto \mathbf{B}'PX \mapsto PX.$$

Therefore, $\phi' \circ Id \circ \phi^{-1} = P$ (left multiplication by P).

DEFINITION 3.12. *The invertible matrix P in both*

$$\mathbf{B} = \mathbf{B}'P \qquad \textit{and} \qquad X' = PX \tag{3.13}$$

is called ***the (transition) matrix of change of basis.***

Certainly, the map sending $X \mapsto PX$ is an isomorphism of F^n onto itself (and such an isomorphism is sometimes called an **automorphism.**)

We can compute the matrix of change of basis explicitly when $V = F^n$ for some positive integer n and the old basis $\mathbf{B}$ is the standard basis $\mathbf{E} = (e_1, ..., e_n)$, but the new basis $\mathbf{B}'$ is arbitrary. In this nice situation, P is simply the inverse of $\mathbf{B}'$:

$$P = \mathbf{B}'^{-1}\mathbf{B} = \mathbf{B}'^{-1}\mathbf{E} = \mathbf{B}'^{-1}. \tag{3.14}$$

Now, in the above discussion, the matrix P was determined in terms of two bases $\mathbf{B}$ and $\mathbf{B}'$. Since we have a relation among three invertible matrices, $P, \mathbf{B}$, and $\mathbf{B}'$, we can determine one of them from the other two. Namely, we can turn the discussion around, starting with just one basis $\mathbf{B}$ and an invertible matrix $P \in GL_n(F)$ = the set of all $n \times n$ invertible matrix with coefficients in a field F, we can define a new basis by the formula:

$$\mathbf{B}' = \mathbf{B}P^{-1}.$$

Since $\mathbf{B} = \mathbf{B}'P$, the vectors $v_i \in \mathbf{B} = (v_1, ..., v_n)$ in the old basis are in the span of $\mathbf{B}'$; hence $\mathbf{B}'$ spans V and, since $\mathbf{B}'$ is linearly independent and it consists of n elements, $\mathbf{B}'$ is indeed a basis of V.

This proves:

COROLLARY 3.2. *Let* $\mathbf{B}$ *be a basis of a vector space* V *of dimension* n. *The other bases are the sets of the form* $\mathbf{B}' = \mathbf{B}Q$, *where* $Q \in GL_n(F)$ *is an invertible matrix.* □

Exercises for Section 3.3

EXERCISE 3.16. *Compute the matrix* P *of change of basis in* $\mathbb{R}^2$ *relating the standard basis* $\mathbf{E} = ((1,0)^t, (0,1)^t)$ *to a new basis* $\mathbf{B}' = ((1,3)^t, (2,2)^t)$.

EXERCISE 3.17. *What is the matrix of change of basis, when the old basis is the standard basis* $(e_1, ..., e_n)$ *and the new basis is* $(e_n, e_{n-1}, ..., e_1)$.

EXERCISE 3.18. *Determine the matrix* P *of change of basis when the old basis is* (e_1, e_2) *and the new basis is* $(e_1 - e_2, e_1 + e_2)$.

EXERCISE 3.19.

(1) *Prove that the* $\mathbf{B} = ((1,2,0)^t, (2,1,2)^t, (3,1,1)^t)$ *is a basis of* $\mathbb{R}^3$.
(2) *Compute the coordinate vector of the vector* $(3,2,1)^t$ *with respect to this basis.*
(3) *Let* $\mathbf{B}' = ((0,1,0)^t, (1,0,1)^t, (2,1,0)^t))$. *Find the matrix* P *relating the old basis* $\mathbf{B}$ *to the new basis* $\mathbf{B}'$.

EXERCISE 3.20. *Let* $\mathbf{B}$ *and* $\mathbf{B}'$ *be two bases of the vector space* $\mathbb{R}^n$. *Prove that the matrix* P *of change of basis from the old basis* $\mathbf{B}$ *to the new basis* $\mathbf{B}'$ *is* $P = \mathbf{B}'^{-1}\mathbf{B}$.

EXERCISE 3.21. *Let* F *be a field and* V *the vector space* F^n. *Prove that there exists a one-to-one correspondence between the set* S *of bases of* V *and the set* $GL_n(F)$ *of all invertible matrices of size* $n \times n$ *with entries in* F.

4. Row space, Column space, and Null space

In this section, we are going to study special spaces associated to a given $m \times n$ matrix

$$A = \begin{bmatrix} a_{11} & a_{12} & \cdot & \cdot & \cdot & a_{1n} \\ a_{21} & a_{22} & \cdot & \cdot & \cdot & a_{2n} \\ \cdot & \cdot & & & & \cdot \\ \cdot & \cdot & & & & \cdot \\ \cdot & \cdot & & & & \cdot \\ a_{m1} & a_{m2} & \cdot & \cdot & \cdot & a_{mn} \end{bmatrix}.$$

DEFINITION 3.13. *The subspace of* $\mathbb{R}^n$ *spanned by the row vectors of the matrix* A *is called the* **row space** *of* A*, and the subspace of* $\mathbb{R}^m$ *spanned by the column vectors of the matrix* A *is called the* **column space** *of* A*. The solution space of the homogeneous system of linear equations* $AX = 0$*, which is a subspace of* $\mathbb{R}^n$ *is called the* **null space** *of* A*.*

In this section we will study how they are related to each other. Their relationship under a linear transformation is studied in a later chapter, Chapter 4.

PROPOSITION 3.17. *A system of linear equations* $AX = B$ *is consistent if and only if* B *is in the column space of* A*.*

PROOF. We denote C_i for the ith column vector of A. Then $AX = B$ becomes (via block multiplication of matrices)

$$x_1C_1 + ... + x_nC_n = B,$$

where $X = (x_1, ..., x_n)^t$, and this is what the proposition says. □

PROPOSITION 3.18. *Let* $\{v_1, ..., v_k\}$ *be a basis for the null space of* A*, and* X_0 *a solution of a consistent (non-homogeneous) system of linear equations* $AX = B$ *with* $B \neq 0$*. Then every solution* X *of* $AX = B$ *is a linear combination:*

$$X = X_0 + \alpha_1 v_1 + ... + \alpha_k v_k, \tag{3.15}$$

and conversely, for all choices of scalars $\alpha_1, ..., \alpha_k$, the vector X in this form is a solution of $AX = B$.

PROOF. Let X be a solution of $AX = B$. Then $AX = B$ and $AX_0 = B$ imply

$$A(X - X_0) = 0,$$

and hence $X - X_0$ is in the solution space of A. Thus, (3.15) holds. Similarly, the converse is very easy. □

REMARK 3.7. *The vector X_0 in (3.15) is called a* ***particular solution*** *of $AX = B$, the expression$X_0 + \alpha_1 v_1 + ... + \alpha_k v_k$ in (3.15) is called a* ***general solution*** *of $AX = B$, and the expression $\alpha_1 v_1 + ... + \alpha_k v_k$ in (3.15) is called a general solution of $AX = 0$. Thus with this terminology, we say that the general solution of $AX = B$ is the sum of a particular solution of $AX = B$ and the general solution of $AX = 0$.*

EXAMPLE 3.16. *Let*

$$\begin{array}{rrrrrrcr} x_1 & + 3x_2 & - 2x_3 & & + 2x_5 & & = & 0 \\ 2x_1 & + 6x_2 & - 5x_3 & - 2x_4 & + 4x_5 & - 3x_6 & = & -1 \\ & & 5x_3 & + 10x_4 & & + 15x_6 & = & 5 \\ 2x_1 & + 6x_2 & & + 8x_4 & + 4x_5 & + 18x_6 & = & 6 \end{array}$$

be a non-homogeneous system of linear equations. If one applies elementary row operations to solve the corresponding augmented matrix expression:

$$\begin{bmatrix} 1 & 3 & -2 & 0 & 2 & 0 & 0 \\ 2 & 6 & -5 & -2 & 4 & -3 & -1 \\ 0 & 0 & 5 & 10 & 0 & 15 & 5 \\ 2 & 6 & 0 & 8 & 4 & 18 & 6 \end{bmatrix}$$

one will obtain its general solution in the following form:

$$\begin{bmatrix} x_1 \\ x_2 \\ x_3 \\ x_4 \\ x_5 \\ x_6 \end{bmatrix} = \begin{bmatrix} -3r-4s-2t \\ r \\ -2s \\ s \\ t \\ 1/3 \end{bmatrix}$$

$$= \begin{bmatrix} 0 \\ 0 \\ 0 \\ 0 \\ 0 \\ 1/3 \end{bmatrix} + r\begin{bmatrix} -3 \\ 1 \\ 0 \\ 0 \\ 0 \\ 0 \end{bmatrix} + s\begin{bmatrix} -4 \\ 0 \\ -2 \\ 1 \\ 0 \\ 0 \end{bmatrix} + t\begin{bmatrix} -2 \\ 0 \\ 0 \\ 0 \\ 1 \\ 0 \end{bmatrix},$$

where $r, s,$ and t are parameters. Thus,

$$X_0 = \begin{bmatrix} 0 \\ 0 \\ 0 \\ 0 \\ 0 \\ 1/3 \end{bmatrix}$$

is a particular solution of the system and

$$X = r\begin{bmatrix} -3 \\ 1 \\ 0 \\ 0 \\ 0 \\ 0 \end{bmatrix} + s\begin{bmatrix} -4 \\ 0 \\ -2 \\ 1 \\ 0 \\ 0 \end{bmatrix} + t\begin{bmatrix} -2 \\ 0 \\ 0 \\ 0 \\ 1 \\ 0 \end{bmatrix}$$

is the general solution of the homogeneous system:

$$\begin{array}{rrrrrrcl} x_1 & +\,3x_2 & -\,2x_3 & & +\,2x_5 & & = & 0 \\ 2x_1 & +\,6x_2 & -\,5x_3 & -\,2x_4 & +\,4x_5 & -\,3x_6 & = & 0 \\ & & 5x_3 & +\,10x_4 & & +\,15x_6 & = & 0 \\ 2x_1 & +\,6x_2 & & +\,8x_4 & +\,4x_5 & +\,18x_6 & = & 0 \end{array},$$

moreover, the vectors

$$\begin{bmatrix} -3 \\ 1 \\ 0 \\ 0 \\ 0 \\ 0 \end{bmatrix}, \begin{bmatrix} -4 \\ 0 \\ -2 \\ 1 \\ 0 \\ 0 \end{bmatrix}, \begin{bmatrix} -2 \\ 0 \\ 0 \\ 0 \\ 1 \\ 0 \end{bmatrix}$$

form a basis for the null space of the system $AX = 0$, *where*

$$A = \begin{bmatrix} 1 & 3 & -2 & 0 & 2 & 0 \\ 2 & 6 & -5 & -2 & 4 & -3 \\ 0 & 0 & 5 & 10 & 0 & 15 \\ 2 & 6 & 0 & 8 & 4 & 18 \end{bmatrix}.$$

PROPOSITION 3.19. *Elementary row operations do not change the null space of a matrix.*

PROOF. That is exactly the way why elementary row operations are introduced in linear algebra. □

PROPOSITION 3.20. *Elementary row operations do not change the row space of a matrix.*

PROOF. Since every elementary row operation has a (unique) inverse elementary row operation, we have this result. □

REMARK 3.8. *But, elementary row operations do change the column space.*

EXAMPLE 3.17. *Let* $A = \begin{bmatrix} 1 & 2 \\ 2 & 4 \end{bmatrix}$ *and* $B = \begin{bmatrix} 1 & 2 \\ 0 & 0 \end{bmatrix}$. *The matrix B is obtained from the matrix A by applying an elementary operation. Here the column space of B is spanned by the column vector $(1,0)^t$ whereas the column space of A is spanned by the column vector $(1,2)^t$.*

PROPOSITION 3.21. *Let A a matrix and B be obtained from A by application of finitely many elementary row operations. Then we have:*

- **a:** *A given set of column vectors of A is linearly independent if and only if the corresponding column vectors of B are linearly independent.*
- **b:** *A given set of column vectors of A form a basis for the column space of A if and only if the corresponding column vectors of B form a basis for the column space of B.*

PROOF. Consider the two homogeneous system of linear equations:

$$AX = 0 \qquad \text{and} \qquad BX = 0.$$

We know that they have the same solution set. Let $c_1, ..., c_n$ and $c_1', ..., c_n'$ be column vectors of A and B, respectively. Then the two systems have only the same trivial solution if and only if both $x_1c_1 + ... + x_nc_n = 0$ and $x_1c_1' + ... + x_nc_n' = 0$ have the same solution $(0,0,...,0)^t$ for $X = (x_1, ..., x_n)^t$. This proves both part a and part b. □

In particular, we have:

COROLLARY 3.3. *If a matrix R is in a row reduced echelon form, we have the following:*

1. *The row vectors with the leading 1's, i.e., the pivots, form a basis for the row space of R.*
2. *The column vectors with the leading 1's, i.e., the pivots, form a basis for the column space of R.* □

EXAMPLE 3.18. *Let*

$$\begin{aligned} v_1 &= (1,-2,0,3), v_2 = (2,-5,-3,6), v_3 = (0,1,3,0), \\ v_4 &= (2,-1,4,-7), v_5 = (5,-8,1,2) \end{aligned}$$

be vectors in $\mathbb{R}^4$.

(1) *Find a subset of those vectors that form a basis for the space spanned by the given five vectors.*
(2) *Express those vectors not in the basis as a linear combination of the basis vectors.*

Consider the following matrix:

$$A = \begin{bmatrix} 1 & 2 & 0 & 2 & 5 \\ -2 & -5 & 1 & -1 & -8 \\ 0 & -3 & 3 & 4 & 1 \\ 3 & 6 & 0 & -7 & 2 \end{bmatrix},$$

where the ith*-column is the vector* v_i*. Applying elementary row operations to this matrix, we obtain its row redduced echelon matrix*

$$B = \begin{bmatrix} 1 & 0 & 2 & 0 & 1 \\ 0 & 1 & -1 & 0 & 1 \\ 0 & 0 & 0 & 1 & 1 \\ 0 & 0 & 0 & 0 & 0 \end{bmatrix}.$$

Let w_i *be the* i^{th}*-column of this row echelon matrix. Then the leading* 1*'s occur in* $w_1, w_2,$ *and* w_4*; and hence* $\{w_1, w_2, w_4\}$ *is a basis for the column space of* B*. Thus, the vectors* $v_1, v_2,$ *and* v_4 *are a basis for the space spanned by* $v_1, ..., v_5$.

Now, since $w_3 = 2w_1 - w_2$*, we conclude* $v_3 = 2v_1 - v_2$*. For this, we see the last part of the proof of Proposition 3.21. Similarly, since* $w_5 = w_1 + w_2 + w_4$*, we conclude that* $v_5 = v_1 + v_2 + v_4$.

COROLLARY 3.4. *The row space of* A *and the column space of* A *have the same dimension.* □

PROPOSITION 3.22. *Let A be a square matrix of size $n \times n$. Then the following statements are equivalent to each other:*

(a): *A can be reduced to the identity matrix by a sequence of elementary row operations.*
(b): *A is a product of elementary matrices.*
(c): *A is invertible.*
(d): *The system of homogeneous equations $AX = 0$ has only the trivial solution $X = 0$.*
(e): $\det(A) \neq 0$.
(f): *A has rank n.*
(g): *The row vectors of A are linearly independent.*
(h): *The column vectors of A are linearly independent.*

PROOF. Exercise. □

EXAMPLE 3.19. *(MAPLE) The following lines compute a basis of the row space and the column space, and the null space, respectively, of a given matrix. The rank of the given matrix is also computed:*

———————

```
interface(echo=2);
with(linalg):
A := array([[1,2,3],[0,1,2],[0,1,1],[1,3,1]]);
result1:=basis(A,'rowspace');
result2:=basis(A, 'colspace');
result3:=basis(transpose(A), 'colspace');
result4:=nullspace(A);
result5:=kernel(A);
result6:=rank(A);
```

———————

REMARK 3.9. *Regardless of whether we want to compute a basis for the row space or column space of a given matrix, this example shows that the basic command of MAPLE returns us, as a computed result, those linearly independent vectors of the* given *generating set, i.e., in terms of the original vectors. If a Computer Algebra System you use does not have this functionality, then you may have to work on the column space of the transpose of the matrix.*

Exercises for Section 3.4

EXERCISE 3.22. *Find a basis for the subspace of $\mathbb{R}^4$ spanned by the given vectors.*

(1) $(1,1,-4,-3)^t, (2,0,2,-2)^t, (2,-1,3,2)^t$.
(2) $(1,1,0,0), (0,0,1,1), (-2,0,2,2), (0,-3,0,3)$.

EXERCISE 3.23. *Find a subset of the vectors that forms a basis for the space spanned by the vectors; then express each vector that is not in the basis as a linear combination of the basis vectors.*

(1) $v_1 = (1,1,-4,-3)^t, v_2 = (2,0,2,-2)^t, v_3 = (2,-1,3,2)^t$.
(2) $v_1 = (1,1,0,0), v_2 = (0,0,1,1), v_3 = (-2,0,2,2), v_4 = (0,-3,0,3)$.

5. Direct Sums

Let V be a vector space, and let $W_1, ..., W_n$ be subspaces of V. In what follows we are going to introduce two concepts for subspaces which are similar to spans and linear independence of vectors.

DEFINITION 3.14. *The set of vectors*

$$W_1 + ... + W_n := \{v \in V : v = w_1 + ... + w_n, \ \text{each } w_i \in W_i\}$$

*is called the **sum** of the subspaces $W_1, ..., W_n$, or their **span**.*

LEMMA 3.4. *The sum is a subspace of V.* □

Clearly, the sum is the smallest subspace containing W_1, ..., and W_n.

DEFINITION 3.15. *The subspaces $W_1, ..., W_n$ are called **independent** if no sum $w_1 + ... + w_n$ with each $w_i \in W_i$ is zero, except for the trivial sum where each $w_i = 0$.*

In other words, the spaces are independent if

$$w_1 + ... + w_n = 0 \qquad \text{and} \qquad w_i \in W_i \qquad \text{implies} \qquad \text{each } w_i = 0.$$

DEFINITION 3.16. *V is the **direct sum** $W_1 \oplus W_2 \oplus ... \oplus W_n$ of $W_1, ..., W_n$ if it is their sum and the W_i are independent.*

LEMMA 3.5. *V is the direct sum of $W_1, ..., W_n$ if and only if every vector $v \in V$ can be written uniquely as*

$$v = w_1 + w_2 + ... + w_n,$$

where each $w_i \in W_i$. $\square$

PROPOSITION 3.23.

- **a:** *A single subspace W of V is independent.*
- **b:** *Two subspaces W_1 and W_2 are independent if and only if $W_1 \cap W_2 = \{0\}$.*

PROOF. Part a is clear. For part b, suppose first that they are independent. Then for $w \in W_1 \cap W_2$, we have $w + (-w) = 0$; hence w must be zero. Conversely, if they are not independent then there exist $w_1 \in W_1$ and $w_2 \in W_2$ such that $w_1 + w_2 = 0$ where at least one of them, say, w_1 is non-zero. Then $w_2 = -w_1 \in W_1 \cap W_2$, so that $W_1 \cap W_2 \neq \{0\}$. $\square$

PROPOSITION 3.24. *Let W_1, W_2 be subspaces of a finite-dimensional vector space V. Then*

$$\dim W_1 + \dim W_2 = \dim(W_1 \cap W_2) + \dim(W_1 + W_2).$$

PROOF. Recall that $W_1 \cap W_2$ is also a subspace of V. Choose a basis $\{u_1, ..., u_r\}$ for $W_1 \cap W_2$ and extend it to a basis of W_1:

$$\{u_1, ..., u_r; x_1, ..., x_{m-r}\}, \tag{3.16}$$

where $r = \dim(W_1 \cap W_2)$ and $m = \dim W_1$. We can do this by Proposition 3.7. Similarly, we can extend it to a basis of W_2:

$$\{u_1, ..., u_r; y_1, ..., y_{n-r}\}, \tag{3.17}$$

where $n = \dim W_2$. The proposition will follow if we prove that the set:

$$\{u_1, ..., u_r; x_1, ..., x_{m-r}; y_1, ..., y_{n-r}\} \tag{3.18}$$

is a basis of the sum $W_1 + W_2$.

(span) For any vector $v \in W_1 + W_2$, it is a sum $v = w_1 + w_2$ with $w_1 \in W_1$ and $w_2 \in W_2$, where we can write w_1 as a linear combination of those vectors in (3.16) and w_2 as a linear combination of those vectors in (3.17); hence v is a linear combination of (3.18).

(linear independence) Suppose that

$$\sum_{i=1}^{r} a_i u_i + \sum_{i=1}^{m-r} b_i x_i + \sum_{i=1}^{n-r} c_i y_i = 0, \qquad \text{with } a_i, b_i, c_i \in F.$$

We abbreviate this as $u + x + y = 0$. Then $y = -u - x \in W_1 \cap W_2$; and hence y is a linear combination, say u' , of $\{u_1, ..., u_r\}$. Thus, $-u' + y = 0$ is a relation among the vectors (3.17), which are linearly independent. Thus it must be a trivial relation; consequently, $y = 0$. Thus our original relation is reduced to $u + x = 0$. Then since (3.16) is a basis, this relation must be also trivial; hence $u = 0$ and $x = 0$. Therefore, the whole relation is trivial, i.e., all the coefficients a_i, b_i, c_i must be zero. □

COROLLARY 3.5. $\dim(W_1 \oplus W_2) = \dim(W_1) + \dim(W_2)$. □

PROPOSITION 3.25. *Let $W_1, ..., W_n$ be subspaces of a finite-dimensional vector space V, and let $\mathbf{B}_i$ be a basis for each W_i.*

- **a:** *The ordered set $\mathbf{B}$ obtained by listing the bases $\mathbf{B}_1, ..., \mathbf{B}_n$ in order is a basis of V if and only if V is the direct sum $W_1 \oplus W_2 \oplus ... \oplus W_n$.*
- **b:** $\dim(W_1 + ... + W_n) \leq (\dim W_1) + ... + (\dim W_n)$, *with equality if and only if the spaces are independent.*

PROOF. The $\Rightarrow$ of part a is very easy. For $\Leftarrow$ of part a, we apply Proposition 3.3. Part b is true by Proposition 3.24. We leave the details as an exercise. □

COROLLARY 3.6. *Let W be a subspace of a finite-dimensional vector space V. Then there exists another subspace W' such that $V = W \oplus W'$.*

PROOF. Let $\{w_1, ..., w_d\}$ be a basis for W. Extend it to a basis

$$\{w_1, ..., w_d; v_1, ..., v_{n-d}\}$$

of V. The span of $\{v_1, ..., v_{n-d}\}$ is the required subspace W' by part a above. □

EXAMPLE 3.20. *(MAPLE) The following lines compute a basis for the sum of the vector spaces spanned by the vectors in { }'s of the command sumbasis:*

———————————

```
interface(echo=2);
with(linalg):
v1:=vector([-1,2,3,0]);
v2:=vector([0,1,2,1]);
v3:=vector([2,-1,-1,1]);
sumbasis( {v1, v2, v3});
sumbasis({v1,v2},{v3});
sumbasis({v2,v1,v3},{v2,v3},{v1,v3});
```

———————————

Exercises for Section 3.5

EXERCISE 3.24. *Prove that the vector space* $\mathbb{R}^{n\times n}$ *of all* $n \times n$ *real matrices is the direct sum of the subspaces of symmetric matrices* $(A = A^t)$ *and of skew-symmetric matrices* $(A = -A^t)$.

EXERCISE 3.25. *Let* W *be the space of* $n \times n$ *matrices whose trace is zero. Find a subspace* W' *so that* $\mathbb{R}^{n\times n} = W \oplus W'$.

EXERCISE 3.26. *Prove that the sum of subspaces is a subspace.*

CHAPTER 4

Linear Transformations

1. The Dimension Formula

We recall that a map T from a vector space V to another vector space W is called a *linear transformation* if

$$(4.1) \qquad T(v_1 + v_2) = T(v_1) + T(v_2) \qquad \text{and} \qquad T(cv) = cT(v),$$

for each $v_1, v_2 \in V$ and for each $c \in F$. Then if T is a linear transformation, we have by induction that

$$(4.2) \qquad T(\sum_i c_i v_i) = \sum_i c_i T(v_i).$$

We already know one important example of a linear transformation: left multiplication by a matrix.

EXAMPLE 4.1. *Let A be an $m \times n$ matrix with entries in F. Then sending a column vector $X \in F^n \longmapsto AX \in F^m$ is clearly a linear transformation from F^n to F^m, called* ***left multiplication*** *by A. This is because $A(X_1 + X_2) = AX_1 + AX_2$ and $A(cX) = cAX$, as matrix multiplications.*

EXAMPLE 4.2. *Let P_n be the set of real polynomial functions of degree $\leq n$ of the form*

$$a_n x^n + a_{n-1} x^{n-1} + \ldots + a_1 x + a_0.$$

Then it is easily checked that P_n is a vector space. Moreover, the derivative $\frac{d}{dx}$ is a linear transformation from P_n to P_{n-1}.

When we are given a linear transformation $T : V \longrightarrow W$, there are two subspaces which are associated with it.

DEFINITION 4.1.

$$\ker T \ : \ = \{v \in V : T(v) = 0\}$$
$$\operatorname{Im} T \ : \ = \{w \in W : w = T(v) \textit{ for some } v \in V\}.$$

$\ker T$ is thus the preimage $T^{-1}(0)$ of the zero vector of W and it is called the **kernel** of T, and $\operatorname{Im} T$ is called the **image** of T as usual.

LEMMA 4.1.

a: $\ker T$ *is a subspace of the domain* V.
b: $\operatorname{Im} T$ *is a subspace of the target* W.

PROOF. We first note that ker T contains 0, since $T(0) = T(0v) = 0T(v) = 0$, and hence they are not empty subsets. Then for $v_1, v_2 \in \ker T$, $T(v_1+v_2) = T(v_1)+T(v_2) = 0+0 = 0$ since T is linear; and hence $v_1+v_2 \in \ker T$. Also, for $v_1 \in \ker T$ and $c \in F$, we have $T(cv_1) = cT(v) = c0 = 0$ since T is linear; hence $cv_1 \in \ker T$. Thus, $\ker T$ is a subspace of V. This proves part a.

For part b, if $w_1, w_2 \in \operatorname{Im} T$, then $w_1 = T(v_1)$ and $w_2 = T(v_2)$ for some v_1 and $v_2 \in V$. Thus, $w_1 + w_2 = T(v_1) + T(v_2) = T(v_1 + v_2)$ and $v_1 + v_2 \in V$. Also, for $c \in F$, $cw_1 = cT(v_1) = T(cv_1)$ and $cv_1 \in V$. Thus, $\operatorname{Im} T$ is a subspace of W. □

LEMMA 4.2. *T is one-to-one if and only if* $\ker T = \{0\}$.

PROOF. Suppose that $T(v_1) = T(v_2)$. Then $0 = T(v_1) - T(v_2) = T(v_1 - v_2)$, thus $v_1 - v_2 \in \ker T = \{0\}$ implies $v_1 = v_2$, and this proves the $\Longleftarrow$ direction. The other direction is obvious. □

REMARK 4.1. *We note that when T is left multiplication by a matrix A, $\ker T$ is just the solution space of the homogeneous linear equation $AX = 0$ and $\operatorname{Im} T$ is just the set of vectors $B \in F^m$ such that the linear equation $AX = B$ has a solution.*

The main result of this section is the **dimension formula:**

THEOREM 4.1. *(Dimension Formula) Let $T : V \longrightarrow W$ be a linear transformation, and assume that V is finite-dimensional. Then*

$$\dim V = \dim(\ker T) + \dim(\operatorname{Im} T).$$

REMARK 4.2. *We will learn in our next course (for example, my 'Abstract Algebra' course) that there is a similar formula (called* ***Counting Formula****), this time in terms of multiplication:*

$$|G| = |H| \cdot [G : H],$$

where H is a subgroup of a group G and $|G|$ is the order of G, that is, the cardinality of G, $[G : H]$ is the number of left cosets of H in G, called the index of H in G.

PROOF. Say that $\dim V = n$. Let $\{u_1, ..., u_k\}$ be a basis for the subspace $\ker T$, and extend it to a basis of V (Proposition 3.7 of Chapter3):

$$\{u_1, ..., u_k; v_1, ..., v_{n-k}\} \tag{4.3}$$

Let $w_i = T(v_i)$ for $i = 1, ..., n-k$. Then by Proposition 3.4, the w_i are all distinct. If we can prove the $S = \{w_1, ..., w_{n-k}\}$ is a basis for $\operatorname{Im} T$, then the formula follows.

Thus, we need to prove that S spans $\operatorname{Im} T$ and that it is linearly independent. For any $w \in \operatorname{Im} T$, $w = T(v)$ for some $v \in V$. We write v as a linear combination of the basis elements:

$$v = a_1u_1 + ... + a_ku_k + b_1v_1 + ... + b_{n-k}v_{n-k},$$

and apply the linear map T:

$$w = 0 + ... + 0 + b_1w_1 + ... + b_{n=k}w_{n-k},$$

since $T(u_i) = 0$. Thus, w is in the span of S. Obviously, $Span(S) \subseteq \operatorname{Im} T$ since $\operatorname{Im} T$ is a subspace of W, and hence $\operatorname{Im} T = Span(S)$.

Now, suppose that a linear relation

$$c_1w_1 + ... + c_{n-k}w_{n-k} = 0 \tag{4.4}$$

is given. We consider $v = c_1v_1 + ... + c_{n-k}v_{n-k}$, where $w_i = T(v_i)$. Thus,

$$T(v) = c_1w_1 + ... + c_{n-k}w_{n-k} = 0.$$

Thus $v \in \ker T$. This means that we may write v in terms of the basis $\{u_1, ...u_k\}$ of $\ker T$,say $v = a_1u_1 + ... + a_ku_k$. Then combining these two expressions for the same v, we get

$$-a_1u_1 + ... + -a_ku_k + c_1v_1 + ... + c_{n-k}v_{n-k} = 0.$$

But, (4.3) is a basis, and hence every coefficients $a_i = 0, c_j = 0$.This shows that the set S is linearly independent, and it completes the proof. □

DEFINITION 4.2. *The dimension of* $\operatorname{Im} T$ *is called the* ***rank*** *of* T*, and the dimension of* $\ker T$ *is called the* ***nullity*** *of* T*.*

Thus, we have proved that

$$\dim V = rank \ + \ nullity.$$

DEFINITION 4.3. *The rank and nullity of an* $m \times n$ *matrix* A *are defined to be the dimensions of the image and kernel of left multiplication by* A*.*

Then the nullity of A is just the dimension of the solution space of the homogeneous linear equation $AX = 0$, and since those vectors B such that $B = AX$ has a solution form the image of left multiplication by A, the rank of A is just the dimension of the subspace spanned by such vectors B. The sum of these dimensions is equal to that of the domain of T.

REMARK 4.3. *Let* B *be a vector in the image of left multiplication by* A*. Then* $AX = B$ *has at least one solution, say* $X = X_0$*. If* $X = X_1$ *is another solution of* $AX = B$*, then their difference* $X_0 - X_1$ *is clearly a solution of the homogeneous equation,* $AX = 0$*; hence* $X_0 - X_1 \in \ker T$*. In other words, this exactly says that adding any solution of the* *homogeneous equation* $AX = 0$ *to a particular solution* X_0 *of the inhomogeneous equation* $AX = B$, *we obtain another solution of the inhomogeneous equation* $AX = B$. *Or,* $X_0 + \ker T$ *is the set of all solutions of* $AX = B$.

REMARK 4.4. *Now suppose that* T *is left multiplication by an* $n \times n$ *square matrix* A. *If* $\det A \neq 0$, *then the system of equations* $AX = B$ *has a unique solution for every* B, *that is, left multiplication by* A *is a one-to-one map, and hence* $\ker T = \{0\}$; *hence the nullity of* T *is zero. Then the dimension formula says that the rank of* T *is the same as the dimension of the domain of* T. *This exactly means that the map* T *is onto. Conversely, if* T *is onto, i.e., if the rank of* T *is equal to the dimension of* V, *the domain of* T, *then the dimension formula says that the nullity of* T *is zero, and hence,* T *is one-to-one; hence* $\det A \neq 0$ *by Proposition 2.4. Now if* $\det A = 0$, *then* T *is not onto, and thus there exists* B *in the target such that* $AX = B$ *does not have a solution. But those* B *that do have solutions have more than one solution by the above argument we made.*

We in fact proved

PROPOSITION 4.1. *A linear operator* T *on a finite-dimensional vector space* V *is one-to-one if and only if it is onto.* □

Exercises for Section 4.1

EXERCISE 4.1. *Let* T *be left multiplication by the matrix*

$$\begin{bmatrix} 1 & 2 & 1 & -1 & 5 \\ 2 & 0 & 2 & 0 & 1 \\ 1 & -1 & 1 & 3 & 2 \\ 0 & 3 & -3 & 2 & 6 \end{bmatrix}.$$

Compute $\ker(T)$ *and* $im(T)$ *explicitly by exhibiting bases of these spaces.*

EXERCISE 4.2. *Determine the rank of the matrix* $\begin{bmatrix} 1 & 2 & 3 & 4 \\ 8 & 7 & 6 & 5 \\ 1 & 3 & 5 & 7 \\ 8 & 6 & 4 & 2 \end{bmatrix}$.

EXERCISE 4.3. *Let A be an $m \times n$ matrix. Prove that the solution space of the homogeneous linear system $AX = 0$ has dimension at least $n - m$.*

EXERCISE 4.4. *Let $(v_1, ..., v_n)$ be an ordered subset of a vector space V. Prove that the map φ: $F^n \longrightarrow V$ defined by $\varphi(X) = v_1x_1 + ... + v_nx_n$ is a linear transformation, where $X = (x_1, ..., x_n)^t$ is a vector in F^n.*

2. The Matrix of a Linear Transformation

We have seen in Example 4.1 that left multiplication by an $m \times n$ matrix A is a linear transformation A: $F^n \longrightarrow F^m$. Conversely, we have:

PROPOSITION 4.2. *Every linear transformation T: $F^n \longrightarrow F^m$ is left multiplication by some $m \times n$ matrix A.*

PROOF. Let T: $F^n \longrightarrow F^m$ be a linear transformation. We may assume that standard bases are used for both F^n and F^m. For each standard basis vector e_j of F^n, we compute its image:

$$T(e_j) = \begin{bmatrix} a_{1j} \\ \cdot \\ \cdot \\ \cdot \\ a_{mj} \end{bmatrix}, \tag{4.5}$$

and we *form* the $m \times n$ matrix

$$A := (a_{ij})$$

having these column vectors as its columns. Then since every column vector X of F^n is written as as a linear combination of the standard basis vectors:

$$X = \begin{bmatrix} x_1 \\ \cdot \\ \cdot \\ \cdot \\ x_n \end{bmatrix} = e_1x_1 + ... + e_nx_n,$$

we get

$$\begin{aligned} T(X) &= \sum_{j=1}^{n} T(e_j)x_j \\ &= \begin{bmatrix} a_{11} \\ \cdot \\ \cdot \\ \cdot \\ a_{m1} \end{bmatrix} x_1 + ... + \begin{bmatrix} a_{1n} \\ \cdot \\ \cdot \\ \cdot \\ a_{mn} \end{bmatrix} x_n = AX = A(X), \end{aligned}$$

for every X; hence $T = A$ as map, and this is what we wanted to prove. □

EXAMPLE 4.3. *The linear map* T: $\mathbb{R}^2 \longrightarrow \mathbb{R}^2$ *sending* $e_1 \mapsto (1,2)^t$ *and* $e_2 \mapsto (1,0)^t$ *is left multiplication by the matrix*

$$\begin{bmatrix} 1 & 1 \\ 2 & 0 \end{bmatrix}$$

If $X = (x_1, x_2)^t = e_1x_1 + e_2x_2$, *then*

$$T(X) = \begin{bmatrix} 1 \\ 2 \end{bmatrix} x_1 + \begin{bmatrix} 1 \\ 0 \end{bmatrix} x_2 = \begin{bmatrix} 1 & 1 \\ 2 & 0 \end{bmatrix} \begin{bmatrix} x_1 \\ x_2 \end{bmatrix} = \begin{bmatrix} x_1 + x_2 \\ 2x_1 \end{bmatrix}.$$

Using the notation established in Section 3 of Chapter 3, where the linear map T was the identity map, we can make a similar computation with an arbitrary linear transformation T: $V \longrightarrow W$, once bases of the two spaces are given. Let $\mathbf{B} = (v_1, ..., v_n)$ and $\mathbf{C} = (w_1, ..., w_m)$ be bases of V and of W, respectively, and consider $T(v_j)$ for each j.

$$T(v_j) = \sum_{i=1}^{m} w_i a_{ij} = w_1 a_{ij} + ... + w_m a_{mj}, \qquad j = 1, 2, ..., n.$$

If we list these n quantities two different ways, we get

$$(T(v_1), ..., T(v_n)) = [w_1, ..., w_m]\, A,$$

where $A = (a_{ij})$ is the matrix whose j^{th} columns are the coordinate vector, $(a_{1j,}a_{2j}, ..., a_{mj})^t$, of $T(v_j)$, for each $j = 1, 2, ..., n$. Thus if we denote the hypervector $(T(v_1), ..., T(v_n))$ by

$$T(\mathbf{B}) = (T(v_1), ..., T(v_n)),$$

$$T(\mathbf{B}) = \mathbf{C}A. \tag{4.6}$$

DEFINITION 4.4. *This $m \times n$ matrix $A = (a_{ij})$ is called the* ***matrix of T with respect to the bases $\mathbf{B}, \mathbf{C}$.***

We note that different choices of the bases will give different matrices.

EXAMPLE 4.4. *In the case that $V = F^n$, $W = F^m$, and the two bases are the standard bases, the matrix A of T is just the matrix constructed in the proof of Proposition 4.2.*

We are now ready to describe all these together. We have a linear transformation T: $(V, \mathbf{B}) \longrightarrow (W, \mathbf{C})$ and isomorphisms ϕ_1: $(V, \mathbf{B}) \longrightarrow F^n$ and ϕ_2: $(W, \mathbf{C}) \longrightarrow F^m$. If we put them together we get

$$\begin{array}{ccc} (V, \mathbf{B}) & \longrightarrow & (W, \mathbf{C}) \\ \phi_1 \downarrow & \circlearrowleft & \downarrow \phi_2 \\ F^n & \dashrightarrow & F^m \end{array} .$$

We want to express the composition $\phi_2 \circ T \circ \phi_1^{-1}$ explicitly: For any column vector $X \in F^n$, $v = \phi_1^{-1}(X)$ is a vector v in V satisfying $v = \mathbf{B}X$; hence $(T \circ \phi_1^{-1})(X) = T(v) = T(\mathbf{B})X = \mathbf{C}AX$. Therefore, $(\phi_2 \circ T \circ \phi_1^{-1})(X) = \phi_2(\mathbf{C}AX) = AX$.We have shown that left multiplication by this matrix A : $X \mapsto AX$ is the map we are looking for $F^n \longrightarrow F^m$, and left multiplication by A makes this diagram commutative, meaning that $A \circ \phi_1 = \phi_2 \circ T$. We summarize this fact as

$$T(\mathbf{B}) = \mathbf{C}A \quad \text{and} \quad Y = AX. \tag{4.7}$$

Thus,

PROPOSITION 4.3. *Any linear transformation between finite-dimensional vector spaces V and W can be identified with matrix multiplication, once bases for the two spaces are chosen.* □

Next, we want to relate the matrix A to another such matrix when we are given a *new* set of bases $\mathbf{B}'$ and $\mathbf{C}'$, $\mathbf{B}'$ for the domain V and $\mathbf{C}'$ for the target W. Under these new bases $\mathbf{B}'$ and $\mathbf{C}'$, we have the following diagram:

$$\begin{array}{ccc} (V,\mathbf{B}) & \longrightarrow & (W,\mathbf{C}) \\ Id\downarrow & \circlearrowleft & \downarrow Id \\ (V,\mathbf{B}') & \longrightarrow & (W,\mathbf{C}') \end{array},$$

where the horizontal arrows are the *same* linear transformation T, and the vertical arrows are the two identity maps, Id: $V \longrightarrow V$, and Id: $W \longrightarrow W$, respectively. Now, we have already seen that the first horizontal arrow gives relations in (4.7). Likewise, the second horizontal arrow gives relations:

$$T(\mathbf{B}') = \mathbf{C}'A' \quad \text{and} \quad Y' = A'X'. \tag{4.8}$$

Then the first and the second vertical arrows give the following relations of expressing the *old* basis vectors as linear combinations of the *new* basis vectors, respectively,

$$\mathbf{B} = \mathbf{B}'P \quad \text{and} \quad \mathbf{C} = \mathbf{C}'Q, \tag{4.9}$$

where P is an $n \times n$ invertible matrix and Q is an $m \times m$ matrix, both completely determined from these last relations, respectively. Therefore, we have

$$\mathbf{C}'A' = T(\mathbf{B}') = T(\mathbf{B}P^{-1}) = T(\mathbf{B})P^{-1} = \mathbf{C}AP^{-1} = \mathbf{C}'QAP^{-1}.$$

Then since $\mathbf{C}'$ is a basis, we finally conclude that

$$A' = QAP^{-1}. \tag{4.10}$$

As before, we have a relation of four quantities; hence if we know three of them, then we know the rest. That is,

PROPOSITION 4.4. *Let A be the matrix of a linear transformation T with respect to some given bases* $\mathbf{B}$ *and* $\mathbf{C}$. *The matrices A' which represent the same T with respect to other bases are those of the form*

$$A' = QAP^{-1}, \tag{4.11}$$

where $Q \in GL_m(F)$ and $P \in GL_n(F)$ are arbitrary invertible matrices.
□

REMARK 4.5. *Once P and Q are chosen in the above expression, the corresponding new bases $\mathbf{B}'$ and $\mathbf{C}'$ are just $\mathbf{B}' = \mathbf{B}P^{-1}$ and $\mathbf{C}' = \mathbf{C}Q^{-1}$, respectively.*

Now given a linear transformation $T\colon V \longrightarrow W$, it is natural to look for bases $\mathbf{B}'$ and $\mathbf{C}'$ of V and W such that the matrix of T with respect to new bases becomes especially nice. In fact, the matrix can be simplified remarkably.

PROPOSITION 4.5.

a: *(Vector space form) Let $T\colon V \longrightarrow W$ be a linear transformation. Bases $\mathbf{B}'$ and $\mathbf{C}'$ can be chosen so that the matrix of T is of the form:*

$$A' = \begin{bmatrix} I_r & \\ & 0 \end{bmatrix}, \tag{4.12}$$

where I_r is the $r \times r$ identity matrix, and $r =$ the rank of T.

b: *(Matrix form) Given any $m \times n$ matrix A, there are matrices $Q \in GL_m(F)$ and $P \in GL_n(F)$ such that QAP^{-1} has the form (4.12).*

c: *The assertions in part a and part b are equivalent.*

PROOF. We first prove part a. Let $(u_1, ..., u_k)$ be a basis of $\ker T$. Extend it to a basis $\mathbf{B}$ for V: $(v_1, ..., v_r; u_1, ..., u_k)$ where $k + r = n$. Let $w_j = T(v_j)$. Then as we have already seen in the proof of Proposition 4.1, $(w_1, ..., w_r)$ is a basis for $\operatorname{Im} T$. Then extend this to a basis $\mathbf{C}$ of W: $(w_1, ..., w_r; x_1, ..., x_s)$. Then the matrix of T with respect to these bases clearly has the form in (4.12).

Now, we prove part c, the equivalence of part a and part b. For part b $\Longrightarrow$ part a, if the matrix A in part a of T originally with respect to some bases $\mathbf{B}, \mathbf{C}$ is not of the form in (4.12), then take $\mathbf{B}' = \mathbf{B}P^{-1}$ and $\mathbf{C}' = \mathbf{C}Q^{-1}$. Then by part b, *with respect to* these new bases, QAP^{-1} has the required form in (4.12). For part a $\Longrightarrow$ part b, we view an arbitrary matrix A as the matrix of the linear transformation, "left multiplication by A", in part a, *with respect to* the standard bases. Then the assertion of part a and (4.11) guarantee the existence of P, Q such that QAP^{-1} has the required form in (4.12). □

REMARK 4.6. *Part b of the above proposition may be proved using only elementary row operations: For an $m \times n$ matrix A, apply elementary row operations $E_1, ..., E_k$ successively to obtain its row reduced echelon form $B_1 = E_k \cdots E_1 A$. Then take its transpose B_1^t and apply elementary row operations $C_1, ..., C_l$ until to get its row reduced echelon form $B_2 = C_l \cdots C_1 B_1^t$; then take its transpose B_2^t.Then $B_2^t = B_1 C_1^t \cdots C_l^t = E_k \cdots E_1 A C_1^t \cdots C_l^t = PAQ^{-1}$ must be the form in (4.12). We note that this proof is constructive.*

DEFINITION 4.5. *Given an $m \times n$ matrix A, define a linear transformation: T: $(V, \mathbf{B}) \longrightarrow (W, \mathbf{C})$ as follows: For any $v \in V$, we have $v = \mathbf{B}X$. Thus, we define*

$$T(v) = T(\mathbf{B})X,$$

where $T(\mathbf{B})$ is defined as $\mathbf{C}A$.

PROPOSITION 4.6. *T: $V \longrightarrow W$ defined above is a linear transformation of which matrix with respect to $\mathbf{B}, \mathbf{C}$ is A.* □

EXAMPLE 4.5. *(MAPLE) The following lines compute a basis of the kernel of the linear transformation by left multiplication by a given matrix:*

```
interface(echo=2);
with(linalg):
A := array([[1,2,3],[0,1,2],[0,1,1],[1,3,1]]);
result4:=kernel(A);
```

Exercises for Section 4.2

EXERCISE 4.5. *Let $T : \mathbb{R}^3 \longrightarrow \mathbb{R}^2$ be the linear transformation defined by the rule $T((x_1, x_2, x_3)^t) = (x_1 + x_2, 2x_3 - x_1)^t$. What is the image of T with respect to the standard bases for both $\mathbb{R}^3$ and $\mathbb{R}^2$?*

EXERCISE 4.6. *Prove that A and A^t have the same rank.*

EXERCISE 4.7. *Let ρ_θ be the translation $R^2 \longrightarrow \mathbb{R}^2$ which rotates the plane counterclockwise about the origin through the angle θ with respect to the standard basis for R^2.*

(1) *Prove that ρ_θ is a linear transformation.*
(2) *Determine the matrix A representing the linear transformation ρ_θ.*
(3) *Derive the addition formulas for the trigonometric functions* cos *and* sin*:*

$$\begin{aligned}\cos(\phi + \theta) &= \cos\phi\cos\theta - \sin\phi\sin\theta \\ \sin(\phi + \theta) &= \sin\phi\cos\theta + \cos\phi\sin\theta.\end{aligned}$$

EXERCISE 4.8. *Let T_1 and T_2 be linear transformations from V to W. Define $T_1 + T_2$ and cT by the rule $(T_1 + T_2)(v) = T_1(v) + T_2(v)$ and $(cT)(v) = c(T(v))$.*

(1) *Prove that $T_1 + T_2$ and cT_1 are linear transformations, and describe their matrices in terms of the matrices for T_1 and T_2.*

(2) *Let L be the set of all linear transformations from V to W. Prove that these laws make L into a vector space.*

3. Linear Operators and Eigenvectors

DEFINITION 4.6. *A linear transformation $T\colon V \longrightarrow V$ of a vector space to itself is called a* **linear operator** *on V.*

Left multiplication by an $n \times n$ matrix with entries in F defines a linear operator on the space F^n of column vectors.

EXAMPLE 4.6. *A rotation ρ_θ of the plane through an angle θ is a linear operator on $\mathbb{R}^2$, whose matrix with respect to the standard basis is*

$$R = \begin{bmatrix} \cos\theta & -\sin\theta \\ \sin\theta & \cos\theta \end{bmatrix}.$$

To verify that R represents a rotation, we write a column vector $X \in \mathbb{R}^2$ in polar coordinates, as $X = (r, \alpha)$. Then in rectangular coordinates, $X = (r\cos\alpha, r\sin\alpha)^t$. Then $RX = (r\cos(\alpha+\theta), r\sin(\alpha+\theta))^t$ by the addition formulas for sine and cosine. Thus in polar coordinates, $RX = (r, \alpha+\theta)$, showing that RX is obtained from X by rotating through the angle θ.

When we talk about a linear operator T, the domain and the target are the same; hence we usually choose a basis $\mathbf{B}$ for both the domain and the target. If we only replace $(W, \mathbf{C})$ with $(V, \mathbf{B})$ in the discussion we have made in Section 2 of this chapter, everything goes in the same way:

$$T(\mathbf{B}) = \mathbf{B}A \qquad \text{and} \qquad Y = AX.$$

This defines the matrix $A = (a_{ij})$ of T with respect to a basis $\mathbf{B}$. With respect to a different basis $\mathbf{B}'$, we have:

$$T(\mathbf{B}') = \mathbf{B}'A' \qquad \text{and} \qquad Y' = A'X'.$$

Then as we have seen in Section 2, the matrices A and A' are related:

$$A' = PAP^{-1},$$

where the invertible matrix P is determined from the change of bases:

$$\mathbf{B} = \mathbf{B}'P.$$

PROPOSITION 4.7. *Let A be the matrix of a linear operator T with respect to a basis $\mathbf{B}$. The matrices A' which represents T for different bases are those of the form*

$$A' = PAP^{-1},$$

for arbitrary $P \in GL_n(F)$. □

REMARK 4.7. *We note that this result may not be proved only using elementary row operations, since we can not apply freely the second phase of transposing the row reduced echelon matrix and applying again elementary row operations, etc., as we have explained in Example 4.6, because we have here only one basis and hence only one change of basis, P, to work with.*

DEFINITION 4.7. *A square matrix A is **similar** to A' if $A' = PAP^{-1}$ for some $P \in GL_n(F)$.*

Before we study eigenvectors and eigenvalues of a linear operator T, we first discuss invariant subspaces under T. We do this because the main tools for analyzing linear operators are the concepts of eigenvector and invariant subspace.

DEFINITION 4.8. *Let $T\colon V \longrightarrow V$ be a linear operator on a vector space V. A subspace W of V is called an **invariant subspace** or a T-invariant subspace if it is carried to itself by the operator:*

$$T(W) \subseteq W. \tag{4.13}$$

When this is so, T defines a linear operator on W, called the **restriction** of T to W.

LEMMA 4.3. *Let W be a k dimensional T-invariant subspace of a finite-dimensional vector space V. Then there exists an ordered basis* $\mathbf{B}$

of V such that the matrix of T with respect to $\mathbf{B}$ is of the form

$$M = \begin{bmatrix} A & B \\ 0 & D \end{bmatrix},$$

where A is a $k \times k$ matrix, and it is the matrix of the restriction of T to W.

PROOF. Let $(w_1, ..., w_k)$ be a basis of W; extend it to a basis of V:

$$\mathbf{B} = (w_1, ..., w_k; v_1, ..., v_{n-k}).$$

Then from the relation, $T(\mathbf{B}) = \mathbf{B}M$, we see that each $T(w_j) \in W$ is just a linear combination of the w_i's only; hence the matrix M looks like the one claimed. Then A is automatically the matrix of the restriction of T to W. □

LEMMA 4.4. *The matrix of T of $V = W_1 \oplus W_2$ of two T-invariant subspaces with each $\mathbf{B}_i$ a basis of W_i is of the form:*

$$M = \begin{bmatrix} A_1 & 0 \\ 0 & A_2 \end{bmatrix},$$

where A_i is the matrix of T restricted to W_i.

PROOF. By Proposition 3.25 of Chapter 3, we obtain a basis $\mathbf{B}$ of V by successively listing elements of $\mathbf{B}_1$ and $\mathbf{B}_2$; and then the matrix M of T is of the required form. □

Traditionally,

DEFINITION 4.9. *An **eigenvector** is defined as a non-zero vector v satisfying*

$$T(v) = cv$$

*for some scalar $c \in F$ and such a scalar is called an **eigenvalue** of a linear operator T associated to the eigenvector v.*

Thus, if $V = \mathbb{R}^n$, then a real eigenvector, if ever exists, is a non-zero vector v such that v and $T(v)$ are parallel. Thus, this definition is more or less geometrically oriented, because it does not tell us how to find such eigenvectors.

Let W be the 1-dimensional subspace spanned by an eigenvector v for a linear operator T. Then W is clearly T-invariant. Conversely, if a one-dimensional subspace W is T-invariant, then each non-zero vector in W is obviously an eigenvector of T. Therefore, an eigenvector can be described as a basis of a one-dimensional T-invariant subspace.

If v is an eigenvector, we know we can extend it to a basis $(v = v_1, ..., v_n)$ of V to obtain the matrix of T which looks like

$$\begin{bmatrix} c & B \\ 0 & D \end{bmatrix},$$

where the scalar c is the eigenvalue associated to v_1; this is exactly the block decomposition in Lemma 4.3 in the case of an invariant subspace of dimension one. This just demonstrates how T-invariant subspaces of dimension one are closely related eigenvectors.

REMARK 4.8. *In the next section, we are going to define an eigenvector in such a way that the definition itself tells us how to compute it, if ever exists.*

Let T: $V \longrightarrow V$ be a linear operator on an n-dimensional vector space V with a basis $\mathbf{B}$. Then we associate an $n \times n$ matrix A which sends the coordinate vector X of any vector v, i.e., $v = \mathbf{B}X$, with respect to $\mathbf{B}$ to AX, which is the coordinate vector of $T(v)$ with respect to the same basis $\mathbf{B}$. Under this setting, we therefore have for any vector v: $T(v) = cv \iff T(\mathbf{B}X) = \mathbf{B}cX$, i.e., $\mathbf{B}AX = T(\mathbf{B})X = \mathbf{B}cX$, that is, $AX = cX$. This definitely allows us to define

DEFINITION 4.10. *An* ***eigenvector*** *of an $n \times n$ matrix A is a non-zero column vector X satisfying*

$$AX = cX$$

for some scalar $c \in F$.

Then, v is an eigenvector of a linear operator T if and only if its coordinate vector X *with respect to any basis* $\mathbf{B}$ is an eigenvector for the associated $n \times n$ matrix A of T, and if this is the case, then their eigenvalues must be exactly the same. In the next section, we will go through this definition to define eigenvectors and eigenvalues that is more suitable to their computation.

COROLLARY 4.1. *Similar matrices have the same eigenvalues.*

PROOF. For, similar matrices represent the *same* linear operator. □

LEMMA 4.5. *Let A be the matrix of a linear operator T on V with respect to some basis $\mathbf{B} = (v_1, ..., v_n)$ of V. Then the basis vector v_j is an eigenvector of T if and only if the jth column of the matrix A has the form $c_j e_j$, where e_j is the unit vector and c_j is a scalar.*

PROOF. This is obvious, since the matrix A is defined by the property $T(v_j) = v_1 a_{1j} + ... + v_n a_{nj}$, for $j = 1, 2, ..., n$. □

COROLLARY 4.2. *The matrix A is a diagonal matrix if and only if every basis vector v_j of $\mathbf{B}$ is an eigenvector.* □

COROLLARY 4.3. *The matrix A of a linear operator is similar to a diagonal matrix if and only if there exists a basis $\mathbf{B}' = (v_1', ..., v_n')$ of V consisting of eigenvectors.* □

We will see in the next section that every linear operator on a finite-dimensional *complex* vector space has at least one eigenvector, and in a later section that in most cases the eigenvectors form a basis. But, a linear operator on a *real* vector space need not have an eigenvector.

EXAMPLE 4.7. *the rotation ρ_θ of the plane does not carry any vector to a parallel one, unless $\theta = 0$ or π. That is, ρ_θ has eigenvectors only when $\theta = 0$ or π.*

EXAMPLE 4.8. *(MAPLE) The following lines compute eigenvalues and eigenvectors of a given matrix:*

———————

```
interface(echo=2);
with(linalg):
A:=matrix([[1,2,3],[4,8,12],[1,-1,0]]);
eigenvals(A);
eigenvects(A);
```

———————

Exercises for Section 4.3

EXERCISE 4.9. *Let $T\colon V \longrightarrow V$ be a linear operator on a real vector space V, and let $c \in \mathbb{R}$. Let E_c be the set of eigenvectors of T with eigenvalue c. Prove that E_c is T-invariant subspace of V.*

EXERCISE 4.10. *Find all invariant subspaces of the real linear operator whose matrix is as follows:*

(1) $\begin{bmatrix} 1 & 1 \\ & 1 \end{bmatrix}$,

(2) $\begin{bmatrix} 1 & & \\ & 2 & \\ & & 3 \end{bmatrix}$.

EXERCISE 4.11. *Let T be a linear operator on $\mathbb{R}^2$. Prove that if T carries a line L to itself, then it also carries every line parallel to L to another line parallel to L.*

EXERCISE 4.12. *Prove that the composition $T_1 \circ T_2$ of linear operators on a vector space is a linear operator, and compute its matrix in terms of the matrices of T_1 and T_2.*

4. The Characteristic Polynomial

DEFINITION 4.11.

a: *For a given scalar $c \in F$, a subset*

$$E_c := \{v \in V : T(v) = cv\}$$

of a finite-dimensional vector space with a linear operator T on it, is called an ***eigenspace*** *of T .*

b: *If $E_c \supsetneqq \{0\}$, then c is called an* ***eigenvalue*** *of T and any non-zero vectors in E_c is called an* ***eigenvector*** *of T.*

We note that $cI - T$, where I is the identity map, is also a linear operator on V, and E_c is just the kernel of $cI - T$.

THEOREM 4.2. *The following statements are all equivalent to each other:*

a: *The scalar $c \in F$ is an eigenvalue of T.*
b: $\ker(cI - T) \supsetneqq \{0\}$.
c: *$cI - T$ is not one-to-one.*
d: $\det(cI - A) = 0$, *where A is the matrix of T with respect to some basis.*
e: *c is a root of the polynomial equation,* $\det(tI - A) = 0$ *for unknown t.*

PROOF. Exercise. □

REMARK 4.9. *Thus, by solving the polynomial equation*

$$\det(tI - A) = 0$$

for t, we first compute eigenvalues c; then by solving a system of linear equations

$$AX = cX$$

for X where the scalar c is already computed, we compute the non-zero coordinate vector X of an eigenvector $v = \mathbf{B}X$.

DEFINITION 4.12. *The* ***characteristic polynomial*** *of a linear operator T is the polynomial*

$$p(t) = \det(tI - A),$$

where A is the matrix of T with respect to some basis of V.

EXAMPLE 4.9. *Let* $A = \begin{bmatrix} a & b \\ c & d \end{bmatrix}$ *be an* 2×2 *matrix. Then* $\det(tI - A) = t^2 - (a+d)t + (ad - bc)$. *Therefore, if its discriminant* $D = (a+d)^2 - 4(ad-bc) = (a-d)^2 + 4bc$ *is positive, which is the case if all the entries* a, b, c, *and* d *of the matrix* A *are positive, then* $\det(tI - A) = 0$ *has real roots.*

PROPOSITION 4.8. *The characteristic polynomial of a linear operator* T *does not depend on the choice of a basis.*

PROOF. Suppose that a second basis leads to a matrix $A' = PAP^{-1}$ by Proposition 4.7. Then we have

$$tI - A' = tI - PAP^{-1} = P(tI)P^{-1} - PAP^{-1} = P(tI - A)P^{-1},$$

and hence

$$\begin{aligned} \det(tI - A') &= \det(P(tI - A)P^{-1}) \\ &= (\det P) \cdot (\det(tI - A)) \cdot (\det(P^{-1})) \\ &= (\det P) \cdot (\det P)^{-1} \cdot \det(tI - A) = \det(tI - A). \end{aligned}$$

□

PROPOSITION 4.9. *The characteristic polynomial has the form*

$$p(t) = t^n - (tr\ A)t^{n-1} + \text{ (intermediate terms)} + (-1)^n \det(A),$$

where tr A, the **trace** *of A, is the sum of the diagonal entries:*

$$tr\ A = a_{11} + \ldots + a_{nn}.$$

PROOF. We just expand the polynomial

$$\det(tI - A) = \det \begin{bmatrix} (t - a_{11}) & -a_{12} & \cdots & -a_{1n} \\ -a_{21} & (t - a_{22}) & \cdots & -a_{2n} \\ \cdot & & & \cdot \\ \cdot & & & \cdot \\ \cdot & & & \cdot \\ -a_{n1} & \cdots & \cdots & (t - a_{nn}) \end{bmatrix}$$

in t of degree n. We also note that $p(0) = (-1)^n \det(A)$. □

REMARK 4.10. *By the previous proposition, we see that all coefficients are independent of the basis. Since the characteristic polynomial is only dependent on the linear operator T, we may define the terms* characteristic polynomial, trace, *and* determinant *of a linear operator T to be those obtained using the matrix of T with respect to an arbitrary basis. Because of this reason, the trace and determinant of a linear operator are prototypical examples of* numerical invariants.

PROPOSITION 4.10. *Let T be a linear operator on a finite-dimensional vector space V.*

- **a:** *If V has dimension n, then T has at most n eigenvalues.*
- **b:** *If F is the field of complex numbers and $V \neq 0$, then T has at least one eigenvalue, and hence it has an eigenvector.*

PROOF. Part a is true, since a polynomial of degree n over any field as at most n roots. This will be proved later when we study polynomials over a field. [cf. Proposition 6.1 in my *Abstract Algebra*, p. 265]. Part b is true since the field of complex numbers is algebraically closed. This fact is usually called the **Fundamental Theorem of Algebra**, originally proved by Gauss. □

EXAMPLE 4.10. *(MAPLE) The following lines compute the characteristic polynomial and eigenvalues of a given matrix:*

———————

```
interface(echo=2);
with(linalg):
A:=matrix([[1,2,3],[4,8,12],[1,-1,0]]);
poly:=charpoly(A,'x');
solve(poly,x);
check:=eigenvals(A);
check2:=eigenvects(A);
```

———————

Exercises for Section 4.4

EXERCISE 4.13. *Compute the characteristic polynomials, eigenvalues, and eigenvectors of the following complex matrices:*

(1) $\begin{bmatrix} -2 & 2 \\ -2 & 3 \end{bmatrix}$,

(2) $\begin{bmatrix} 1 & i \\ -i & 1 \end{bmatrix}$.

EXERCISE 4.14. *Prove that the eigenvalues of a real symmetric* 2×2 *matrix are real numbers.*

EXERCISE 4.15. *Prove that a real* 2×2 *matrix whose off-diagonal entries are positive has real eigenvalues.*

EXERCISE 4.16. *Prove that a real* 3×3 *matrix has at least one real eigenvalue.*

EXERCISE 4.17. *Determine the characteristic polynomial of the matrix*

$$\begin{bmatrix} 0 & 1 & & & & & \\ 1 & 0 & 1 & & & & \\ & 1 & \cdot & & & & \\ & & 1 & \cdot & & & \\ & & & & \cdot & 1 & \\ & & & & & 0 & 1 \\ & & & & & 1 & 0 \end{bmatrix}.$$

EXERCISE 4.18. *Do* A *and* A^t *have the same eigenvalues? the same eigenvectors?*

EXERCISE 4.19. *Let* T *be a linear operator having two linearly independent eigenvectors with the same eigenvalue.*

(1) *Is it true that* λ *is a multiple root of the characteristic polynomial of* T*?*
(2) *Suppose that* λ *is a multiple root of the characteristic polynomial. Does* T *have two linear independent eigenvector with eigenvalue* λ*?*

5. Orthogonal Matrices and Rotations

We start with a

DEFINITION 4.13. *A real $n \times n$ matrix A is called* ***orthogonal*** *if $A^{-1} = A^t$.*

LEMMA 4.6. *The determinant of an orthogonal matrix A is ± 1.*

PROOF. Since $I = AA^t$, $1 = \det(AA^t) = \det(A)\det(A^t) = (\det(A))^2$. □

To study orthogonal matrices, we need the dot product of vectors.

DEFINITION 4.14. *The* ***dot product*** *or* ***Euclidean inner product*** *of column vectors $X = (x_1, ..., x_n)^t$ and $Y = (y_1, ..., y_n)^t$ in $\mathbb{R}^n$ is*

$$X \cdot Y = x_1 y_1 + ... + x_n y_n.$$

REMARK 4.11. *Obviously, we see that $X \cdot Y = Y \cdot X$, i.e., it is symmetric, and in matrix form*

$$X \cdot Y = X^t Y, \tag{4.14}$$

and this is called the ***formula for dot product.***

The first main property of the dot product of vectors in $\mathbb{R}^2$ and $\mathbb{R}^3$ is that $X \cdot X$ is the square of the length of the vector:

$$|X|^2 = x_1^2 + x_2^2 \qquad \text{or} \qquad x_1^2 + x_2^2 + x_3^2,$$

respectively. This property is a direct consequence of Pythagorean theorem , and it is a basis for the definition of length of vectors in $\mathbb{R}^n$:

DEFINITION 4.15. *The* ***length*** *$|X|$ of a column vector $X = (x_1, ..., x_n)^t$ in $\mathbb{R}^n$ is defined by*

$$|X|^2 = X \cdot X = x_1^2 + ... + x_n^2. \tag{4.15}$$

The ***distance*** *between two column vectors X, Y is defined to be the length $|X - Y|$ of $X - Y$.*

Let Δ be a triangle with side lengths a, b, c where θ is the angle subtended by the sides a and b. Then by Pythagorean theorem,

$$(a\sin\theta)^2 + (b - a\cos\theta)^2 = c^2,$$

and hence using the fact that $\sin^2\theta + \cos^2\theta = 1$, we obtain the **law of cosines**:

$$c^2 = a^2 + b^2 - 2ab\cos\theta.$$

The second important property of dot product in $\mathbb{R}^2$ and $\mathbb{R}^3$ is the formula

LEMMA 4.7.

$$X \cdot Y = |X|\,|Y|\cos\theta, \tag{4.16}$$

where θ is the angle between the vectors.

PROOF. We apply the law of cosines to the triangle with vertices $0, X, Y$. Its sides are then $|X|, |Y|$, and $|X - Y|$, where the angle θ is between sides $|X|$ and $|Y|$. Thus, by the law of cosines, we have

$$X - Y \cdot X - Y = X \cdot X + Y \cdot Y - 2\,|X|\,|Y|\cos\theta.$$

Then since $X - Y \cdot X - Y = X \cdot X - 2X \cdot Y + Y \cdot Y$, we obtain the formula by comparing terms in both sides of the equality. □

The most important application of (4.16) is that it relates the angle and the dot product for non-zero column vectors X and Y in $\mathbb{R}^2$ and $\mathbb{R}^3$:

$$\cos\theta = \frac{X \cdot Y}{|X|\,|Y|},$$

and hence $\left|\frac{X \cdot Y}{|X||Y|}\right| \leq 1$ for non-zero column vectors X, Y in $\mathbb{R}^2$ and $\mathbb{R}^3$. Therefore, in order for this to be defined as the angle, i.e., $\cos\theta$, of two non-zero column vectors $X, Y \in \mathbb{R}^n$ for a general $n = 2, 3, \ldots$ we need to prove

LEMMA 4.8. *(Cauchy-Schwarz Inequality) For column vectors $X, Y \in \mathbb{R}^n$,*

$$|X \cdot Y| \leq |X|\,|Y|. \tag{4.17}$$

PROOF. If $X = 0$, then both sides are zero. Assume now that $X \neq 0$. For any real number t, we have

$$0 \leq (tX + Y) \cdot (tX + Y) = (X \cdot X)t^2 + 2(X \cdot Y)t + Y \cdot Y,$$

which is about a quadratic polynomial in t; hence its discriminant must satisfy:

$$(2(X \cdot Y))^2 - 4(X \cdot X)(Y \cdot Y) \leq 0.$$

From this inequality, we obtain the required inequality. □

On the other hand, $\cos\theta$ is monotonically decreasing when $0 \leq \theta \leq \pi$. Thus, for each $\frac{X \cdot Y}{|X||Y|}$, there exists a *unique* angle θ such that $\cos\theta$ equals $\frac{X \cdot Y}{|X||Y|}$.

DEFINITION 4.16. *We define "this angle" θ to be the* ***angle*** *between two non-zero column vectors $X, Y \in \mathbb{R}^n$. That is,*

$$\cos\theta = \frac{X \cdot Y}{|X|\,|Y|}.$$

DEFINITION 4.17. *Let X, Y be two non-zero column vectors in $\mathbb{R}^n$. X is* ***orthogonal*** *to Y if $X \cdot Y = 0$.*

Using orthogonality, we can characterize orthogonal matrices more clearly:

PROPOSITION 4.11. *The following statements on a real* invertible $n \times n$ *matrix A are equivalent to each other:*

a: *A is orthogonal.*

b: *Multiplication by A preserves dot product, that is,*

$$AX \cdot AY = X \cdot Y$$

for each column vectors X, Y.

c: *The columns of A are mutually orthogonal unit vectors.*

PROOF. We write $X \cdot Y = X^tY$. If A is orthogonal, $A^tA = I$. Thus

$$X \cdot Y = X^tY = X^tA^tAY = (AX)^tAY = AX \cdot AY.$$

Conversely, suppose that $X^tY = X^tA^tAY$ for all X and Y, i.e., $X^t(I - A^tA)Y = 0$ for all X and all Y. Then since

$$e_i^t B e_j = b_{ij},$$

for any matrix $B = (b_{ij})$, in particular, taking $X = e_i$ and $Y = e_j$, we see that $e_i^t(I - A^tA)e_j = 0$ for all i, j implies $I - A^tA = 0$.This proves the equivalence of part a and part b.

To prove part a and part c are equivalent, let A_j be the jth column of the matrix A. Then $A^tA = [A_1^t, ..., A_n^t]^t [A_1, ..., A_n]$; hence, the (i, j)-th entry of the $n \times n$ matrix A^tA is $A_i^tA_j = A_i \cdot A_j$. Therefore, $A^tA = I$ if and only if $A_i \cdot A_i = 1$ for all i, and $A_i \cdot A_j = 0$ for all $i \neq j$, which is to say that the columns of A have length 1 and are mutually orthogonal. □

We note that if the columns of A are mutually orthogonal unit vectors, then these columns as column vectors are necessarily linearly independent, and hence they are a basis of $\mathbb{R}^n$. (Cf. Proposition 3.15.)

DEFINITION 4.18. *A basis consisting of mutually orthogonal unit vectors is called an* ***orthonormal*** *basis.*

Thus, an orthogonal matrix is one whose columns form an orthonormal basis.

DEFINITION 4.19. *Left multiplication by an orthogonal matrix is also called an* ***orthogonal operator****.*

Thus, the orthogonal operators on $\mathbb{R}^n$ are those that preserve dot product. Thus, orthogonal operators on $\mathbb{R}^n$ preserves lengths, triangles, angles, and shapes in general.

The geometric meaning of multiplication by an orthogonal matrix can be explained in terms of rigid motions.

DEFINITION 4.20. *A* ***rigid motion*** *or* ***isometry*** *of $\mathbb{R}^n$ is a map m: $\mathbb{R}^n \longrightarrow \mathbb{R}^n$ which is distance preserving, that is, it is a map satisfying the* *following condition: If X, Y are two points of $\mathbb{R}^n$, then the distance from X to Y is equal to that from $m(X)$ to $m(Y)$:*

$$|m(X) - m(Y)| = |X - Y|. \tag{4.18}$$

Such a rigid motion carries a triangle to a congruent triangle, and therefore it preserves angles and shapes in general.

DEFINITION 4.21. *The set of all rigid motions of $\mathbb{R}^n$ is called the* ***group M_n of motions.***

LEMMA 4.9. *The only map $\mathbb{R}^n \longrightarrow \mathbb{R}^n$ that preserves dot product* and *that also fixes each of the standard basis vectors e_i is the identity map.*

PROOF. We are going to use the shorthand $'$ to denote the map m, writing $m(X) = X'$.

If m preserves dot product, then $X \cdot e_j = X' \cdot e_j'$ for any column vector $X = (x_1, ..., x_n)^t$. If moreover $e_j = e_j'$ as well, then

$$x_j = X \cdot e_j = X' \cdot e_j' = X' \cdot e_j = x_j',$$

for all j, where $X' = (x_1', ..., x_n')^t$. Therefore, $X = X'$, and m is the identity map. □

PROPOSITION 4.12. *Let m: $\mathbb{R}^n \longrightarrow \mathbb{R}^n$. Then the following statements on the map m are equivalent to each other:*

a: *m is a rigid motion which fixes the origin.*

b: *m preserves dot product; that is, for all $X, Y \in \mathbb{R}^n$,*

$$m(X) \cdot m(Y) = X \cdot Y.$$

c: *m is left multiplication by an orthogonal matrix.*

PROOF. We are again going to use the shorthand $'$ to denote the map m, writing $m(X) = X'$.

(part a $\Rightarrow$ part b) Suppose that m is a rigid motion fixing the origin. Since m preserves distance,

$$(X' - Y') \cdot (X' - Y') = (X - Y) \cdot (X - Y) \tag{4.19}$$

for all column vectors X, Y. On the other hand, since it fixes the origin, setting $Y = 0$ shows that $X' \cdot X' = X \cdot X$ for all X. Thus, using this, we expand both sides of (4.19) to obtain $X' \cdot Y' = X \cdot Y$, and this proves part a implies part b.

(part b $\Rightarrow$ part c) Now suppose that m preserves dot product. Then the images $e_1', ..., e_n'$ of the standard basis vectors e_i's are again orthonormal; hence they form another basis $\mathbf{B}'$ of $\mathbb{R}^n$ by Proposition 4.11. Let $A = [e_1', ..., e_n']$ be the matrix whose columns are the e_i''s. We thus have left multiplication by A: $\mathbb{R}^n \longrightarrow \mathbb{R}^n$. Our strategy to prove part c is to prove that

$$A^{-1} \circ m = Id.$$

For this we go as follows: By Proposition 4.11, the matrix A is orthogonal; hence its inverse A^{-1} is also orthogonal. This is because $(A^{-1})^t = (A^t)^{-1} = A = (A^{-1})^{-1}$. Thus, left multiplication by A^{-1} preserves dot product too. Thus, the composed map $A^{-1} \circ m$ preserves dot product. Moreover, because we have

$$\left(A^{-1} \circ m\right)(e_i) = A^{-1}e_i' = [x_1, ..., x_n]^t$$

for each i, we see that

$$e_i' = A\,[x_1, ..., x_n]^t = [e_1', ..., e_n']\,[x_1, ..., x_n]^t = e_1' x_1 + ... + e_n' x_n$$

for each i. Then since the e_i' is a basis, this means that $x_i = 1$ and $x_j = 0$ if $j \neq i$, for each i. Thus,

$$\left(A^{-1} \circ m\right)(e_i) = A^{-1}e_i' = e_i,$$

for each i, that is, it fixes each standard basis vector e_i. Therefore, by Lemma 4.9, we conclude that $A^{-1} \circ m$ is indeed the identity map; consequently, m is left multiplication by A, and this proves part c from part b.

(part c $\Rightarrow$ part a) Finally, suppose that m is a linear operator whose matrix A is orthogonal. Then since m is linear, $|X' - Y'| = |(X - Y)'| = |X - Y|$, by Proposition 4.11. Thus m is a rigid motion. Then since any linear operator fixes the origin, this proves that part c implies part a. $\square$

COROLLARY 4.4. *A rigid motion which fixes the origin is a linear operator.* □

One class of rigid motions which do not fix the origin, and hence are not linear operators, is the translations.

DEFINITION 4.22. *Given a fixed vector* $b = (b_1, ..., b_n)^t$ *in* $\mathbb{R}^n$, ***translation*** *by* b *is the map*

$$t_b(X) = X + b. \tag{4.20}$$

Any translation is obviously a rigid motion without any rotation at all. Thus a rotation and a translation are two ingredients of a rigid motion:

PROPOSITION 4.13. *Every rigid motion* m *is the composition of an orthogonal linear operator and a translation. In other words, it has the form* $m(X) = AX + b$ *for some uniquely determined orthogonal matrix* A *and some uniquely determined vector* b.

PROOF. Let $b = m(0)$. Then $t_{-b}(b) = 0$; hence the composed operator $t_{-b} \circ m$ is a rigid motion that fixes the origin. Thus, by Proposition 4.12 it is left multiplication by an orthogonal matrix A: $(t_{-b} \circ m)(X) = AX$. Thus, $(t_b\circ\ (t_{-b} \circ m))(X) = AX + b$, i.e., $m(X) = AX + b$.

We note that both b and A are uniquely determined because $b = m(0)$ and A is the operator $t_{-b} \circ m$. □

We already know that the determinant of an orthogonal matrix is ± 1.

DEFINITION 4.23. *An orthogonal operator is said to be* ***orientation-preserving*** *if its determinant is* $+1$, *and* ***orientation-reversing*** *if its determinant is* -1. *Similarly, a rigid motion* $m(X) = AX+b$ *is orientation-preserving if* $\det A = +1$, *and orientation-reversing if* $\det A = -1$.

EXAMPLE 4.11. *A motion of* $\mathbb{R}^2$ *is orientation-reversing if it flips the plane over, and orientation-preserving if it does not.*

EXAMPLE 4.12. *(MAPLE) The following lines compute the inner product, i.e., the dot product, of given two vectors:*

———————

```
interface(echo=2);
with(linalg):
v1:=vector([-1,2,3,0]);
v2:=vector([0,1,2,1]);
innerprod(v1,v2);
```

———————

EXAMPLE 4.13. *(MAPLE) The following lines decides whether a given matrix is an orthogonal matrix or not, based on Proposition 4.11:*

———————

```
interface(echo=2);
with(linalg):
A:=matrix([[1,2,3],[4,8,12],[1,-1,0]]);
orthog(A);
B := array( [[-1/2, sqrt(3)/2], [sqrt(3)/2, 1/2]] );
orthog(B);
```

———————

Exercises for Section 4.5

EXERCISE 4.20. *Prove that every orthonormal set of n vectors in $\mathbb{R}^n$ is a basis.*

EXERCISE 4.21. *What is the matrix of the three-dimensional rotation through the angle θ about the axis e_2?*

EXERCISE 4.22. *Prove that a rigid motion is bijective.*

6. Diagonalization

We are going to prove in this section that for "most" linear operators on a complex vector space, there exists a basis such that the matrix of the operator is a diagonal matrix. We recall that every complex polynomial of positive degree n has a root (cf. Proposition 5.6), and hence every linear operator T on a complex vector space V has an eigenvector.

PROPOSITION 4.14.

a: *(Vector space form) Let T be a linear operator on a finite-dimensional complex vector space V. Then there exists a basis $\mathbf{B}$ of V such that the matrix A of T is upper triangular.*

b: *(Matrix form) Every complex $n \times n$ matrix A is similar to an upper triangular matrix. In other words, there exists an invertible matrix $P \in GL_n(\mathbb{C})$ such that PAP^{-1} is upper triangular.*

PROOF. By Proposition 4.7, the two assertions are equivalent. Thus, we are going to prove part b.

Since the characteristic equation, $\det(tI - A) = 0$ has at least one root, we have an eigenvector, call it v_1'. Extend it to a basis $\mathbf{B}' = (v_1', ..., v_n')$ for V. By Lemma 4.5, the first column of the matrix A' of T with respect to $\mathbf{B}'$ is $(c_1, 0, ..., 0)^t$, where c_1 is the eigenvalue of v_1'. Thus, A' looks like

$$A' = \begin{bmatrix} c_1 & * & \cdots & * \\ 0 & & & \\ \cdot & & & \\ \cdot & & B & \\ \cdot & & & \\ 0 & & & \end{bmatrix}, \tag{4.21}$$

where B is an $(n-1) \times (n-1)$ matrix. This exactly means that there exists an invertible matrix $P \in GL_n(\mathbf{C})$ such that $A' = PAP^{-1}$ has the above form. To prove part b, we apply induction on n. By induction hypothesis, we may assume that there exists an $(n-1) \times (n-1)$ invertible matrix $Q \in GL_{n-1}(\mathbf{C})$ such that QBQ^{-1} is upper triangular has been proved. We define

$$Q_1 = \begin{bmatrix} 1 & 0 & \cdots & 0 \\ 0 & & & \\ \cdot & & & \\ \cdot & & Q & \\ \cdot & & & \\ 0 & & & \end{bmatrix}.$$

Then Q_1 is obviously invertible and hence so is Q_1P, and we see that

$$(Q_1P)A(Q_1P)^{-1} = Q_1(PAP^{-1})Q_1^{-1} = Q_1A'Q_1^{-1},$$

and (via block multiplication of matrices)

$$Q_1A'Q_1^{-1} = \begin{bmatrix} c_1 & * & \cdots & * \\ 0 & & & \\ \cdot & & & \\ \cdot & & QBQ^{-1} & \\ \cdot & & & \\ 0 & & & \end{bmatrix},$$

is again upper triangular. □

REMARK 4.12. *The reason why the above proof works is that every complex polynomial has a root (Gauss' Fundamental Theorem of Algebra). The same proof will work for any field F, provided that all the roots of the characteristic polynomial are in the field F.*

COROLLARY 4.5. *Let F be a field.*

a: *Let T be a linear operator on a finite-dimensional vector space V over F. Suppose that the characteristic polynomial of T factors into linear factors in the field F. Then there is a basis* **B** *of V such that the matrix A of T is triangular.*

b: *(Matrix form) Let A be an $n \times n$ matrix of which characteristic polynomial factors into linear factors in the field F. Then there exist a matrix $P \in GL_n(F)$ such that PAP^{-1} is triangular.*

PROOF. The proof is the same as that of the above proposition, but since our field F is may not be algebraically closed like the field $\mathbb{C}$ of complex numbers, in order to make the induction step in the proof of the above proposition work we have to make it sure that the characteristic polynomial of the matrix B in the above proof is

$$p(t)/(t - c_1),$$

where $p(t)$ is the characteristic polynomial of A, $\det(tI - A)$. For this, we note that $p(t)$ is also the characteristic polynomial of $A' = PAP^{-1}$ by Proposition 5.32. Then from Eq. (4.21) we see that

$$p(t) := \det(tI - A) = (t - c_1)\det(tI - B),$$

and hence our hypothesis that the characteristic polynomial factors into linear factors carries over from A to B under the induction step. □

DEFINITION 4.24. *Let A be an $n \times n$ matrix with entries in F. We say that it is* ***diagonalizable*** *if there exists an invertible matrix $P \in GL_n(F)$ such that PAP^{-1} is diagonal matrix.*

PROPOSITION 4.15. *Let A be a an $n \times n$ matrix. The following statements are equivalent to each other:*

a: *(Vector space form) A is diagonalizable.*
b: *A has n linearly independent eigenvectors.*

PROOF. In what follows, we use block multiplication as much as we could. We write a diagonal matrix $D = (d_{ij})$ as $d_{ii} = \lambda_i$ for each i, $A = [A_1, ..., A_n]$ where A_i are columns of A for each i, $P = [P_1, ..., P_n]$ where P_i are columns of P. By Proposition 3.15 we know that P is invertible if and only if its column vectors P_i are non-zero linearly independent vectors. To prove the equivalence our trick is to work with $AP = PD$, instead of $DP = PA$. Obviously, for an invertible P, the condition $AP = PD$ is exactly the same as

$$[AP_1, ..., AP_n] = [P_1\lambda_1, ..., P_n\lambda_n] = [\lambda_1 P_1, ..., \lambda_n P_n]\,, \tag{4.22}$$

which is then exactly the same as saying that A has exactly n linearly independent eigenvectors P_i. This proves the equivalence. □

REMARK 4.13. *We note that the equivalence of part a and part b of Theorem 5.22 below is a corollary of this proposition.*

In the next chapter, we will learn that *dot product* $X \cdot Y = X^t I Y$ of real column vectors $X, Y \in \mathbb{R}^n$ is a main example of a bilinear, symmetric, and positive definite form on a vector space V. It is symmetric if and only if its associated real identity matrix I is symmetric.

DEFINITION 4.25. *Let A be an $n \times n$ matrix with entries in F. We say that it is* **orthogonally diagonalizable** *if there exists an orthogonal matrix $P \in GL_n(F)$ such that PAP^{-1} is a diagonal matrix.*

Then because of the following result, the meaning of being symmetric for a *real* matrix is very special:

THEOREM 4.3. *Let A be a* real *$n \times n$ matrix. The following statements are equivalent to each other.*

- **a:** *A is orthogonally diagonalizable.*
- **b:** *A has an orthonormal set of n eigenvectors.*
- **c:** *A is symmetric.*

PROOF. We recall that the columns of an orthogonal matrix are orthonormal; hence the equivalence of part a and part b follows from Proposition 4.15. Part a implies part c is obvious, because $A = PDP^t = A^t$ if P is an orthogonal matrix and D is a diagonal matrix.

But, the proof of part c $\Longrightarrow$ part b requires either the **spectral theorem** (Theorem 5.7 or Theorem 5.8) or the famous **Schur's lemma**: If a real $n \times n$ matrix has only real eigenvalues, then A is orthogonally similar to an upper triangular matrix U. In our case, a real symmetric $n \times n$ matrix is a hermitian matrix, and we will later prove that eigenvalues of a hermitian matrix are real numbers (Proposition 5.10). Thus, by Schur A is orthogonally similar to an upper triangular matrix U. Then since A is symmetric, U must be diagonal. □

REMARK 4.14. *We describe this theorem here in order to see some reason why we need to study complex vector spaces. We shall see later that the proof of the spectral theorem is almost the same as the proof of Proposition 4.14. We only postpone its proof simply because we do not introduce the needed terminologies for complex vectors spaces yet.*

DEFINITION 4.26. *Let A be an $n \times n$ matrix with entries in F. We say that it is **congruently diagonalizable** if there exists an invertible matrix $P \in GL_n(F)$ such that PAP^t is diagonal matrix.*

It is obvious that if it is orthogonally diagonalizable then it is congruently diagonalizable. But, the problem with this definition is: how easily we could figure out such an invertible matrix P. As the following example shows, even when we are give a real symmetric 3×3 matrix it can be very difficult.

EXAMPLE 4.14. *Let*

$$A = \begin{bmatrix} 1 & 2 & 3 \\ 2 & 4 & 7 \\ 3 & 7 & 7 \end{bmatrix}. \tag{4.23}$$

We note that it is symmetric and it is invertible (but later in Chapter 5 we will see that it is not *positive definite, since the determinant of its submatrix* $\begin{bmatrix} 1 & 2 \\ 2 & 4 \end{bmatrix}$ *is zero (Theorem 5.3)). If we apply elementary row operations except row exchanges, we get*

$$B := E_2E_1AE_1^tE_2^t = \begin{bmatrix} 1 & 0 & 0 \\ 0 & 0 & 1 \\ 0 & 1 & -2 \end{bmatrix},$$

a non-diagonal matrix, where

$$E_1 = \begin{bmatrix} 1 & 0 & 0 \\ 0 & 1 & 0 \\ -3 & 0 & 1 \end{bmatrix}, \ E_2 = \begin{bmatrix} 1 & 0 & 0 \\ -2 & 1 & 0 \\ 0 & 0 & 1 \end{bmatrix}.$$

What else elementary row operations can we apply to make B diagonal? None. This means that invertible matrices of the type, products of elementary matrices, are not sufficient to actually make it congruent to a diagonal matrix.

We recall Corollary 4.2 that the matrix A of a linear operator T is a diagonal matrix if and only if every basis vector v_j of $\mathbf{B}$ is an eigenvector.

PROPOSITION 4.16. *Let $v_1, ..., v_r \in V$ be eigenvectors for a linear operator T, with distinct eigenvalues $c_1, ..., c_r$. Then the set $(v_1, ..., v_r)$ is linearly independent.*

PROOF. Suppose that

$$0 = a_1 v_1 + ... + a_r v_r.$$

We need to show that $a_i = 0$ for each $i = 1, ..., r$. Applying T on both sides, we have

$$0 = a_1 T(v_1) + ... + a_r T(v_r) = a_1 c_1 v_1 + ... + a_r c_r v_r.$$

Since any non-zero vector is linearly independent, we may assume that $c_r \neq 0$. From these two equations we eliminate v_r:

$$0 = a_1(c_r - c_1)v_1 + ... + a_{r-1}(c_r - c_{r-1})v_{r-1}.$$

By induction hypothesis, we may assume that the $v_1, ..., v_{r-1}$ are linearly independent. Thus, the coefficients $a_i(c_r - c_i) = 0$ for $i = 1, ..., r-1$, and hence $a_i = 0$ for $i = 1, ..., r-1$, since the c_i are distinct. Thus, the original relation is reduced to $0 = a_r v_r$; and hence a_r must be zero since $v_r \neq 0$. □

Combining Corollary 4.2 and this proposition we have

THEOREM 4.4. *Let T be a linear operator on a vector space V of dimension n over a field F. Assume that its characteristic polynomial has n distinct roots in F. Then there is a basis for V with respect to which the matrix of T is diagonal.* □

REMARK 4.15. *We note that the diagonal entries are determined, except for their order, by the linear operator T. They are the eigenvalues.*

EXAMPLE 4.15. *Consider the matrix*

$$A = \begin{bmatrix} 3 & 2 \\ 1 & 4 \end{bmatrix}.$$

It is easy to compute its eigenvectors, $v_1 = (1,1)^t$ and $v_2 = (2,-1)^t$. Since they are linearly independent, they form a basis $\mathbf{B} = (v_1, v_2)$ of $\mathbb{R}^2$.

Thus the matrix P *for change of basis from the standard basis* $\mathbf{E}$ *to* $\mathbf{B}$ *is:* $\mathbf{I} = \mathbf{B}P$. *Thus,*

$$P = \mathbf{B}^{-1} = \begin{bmatrix} 1 & 2 \\ 1 & -1 \end{bmatrix}^{-1} = \begin{bmatrix} \frac{1}{3} & \frac{2}{3} \\ \frac{1}{3} & -\frac{1}{3} \end{bmatrix},$$

and

$$PAP^{-1} = \begin{bmatrix} \frac{1}{3} & \frac{2}{3} \\ \frac{1}{3} & -\frac{1}{3} \end{bmatrix} \begin{bmatrix} 3 & 2 \\ 1 & 4 \end{bmatrix} \begin{bmatrix} 1 & 2 \\ 1 & -1 \end{bmatrix} = \begin{bmatrix} 5 & 0 \\ 0 & 2 \end{bmatrix} = A'$$

is indeed diagonal.

COROLLARY 4.6. *If a basis* $\mathbf{B}$ *of eigenvectors of* A *in* F^n *(with the standard basis) is known, then* $A' = PAP^{-1}$ *is diagonal.* □

Here $P = \mathbf{B}^{-1}$.

EXAMPLE 4.16. *Let*

$$A = \begin{bmatrix} 1 & 2 & 3 \\ 2 & 4 & 7 \\ 3 & 7 & 7 \end{bmatrix}.$$

It is a very simple real and symmetric matrix. Since its characteristic polynomial

$$\det \begin{bmatrix} t-1 & -2 & -3 \\ -2 & t-4 & -7 \\ -3 & -7 & t-7 \end{bmatrix} = t^3 - 12t^2 - 23t + 1, \tag{4.24}$$

is irreducible over the field $\mathbb{Q}$ *of rational numbers, we have to compute its eigenvalues numerically, for instance, using the 'fsolve' command of the Computer Algebra System, MAPLE:*

$$\begin{aligned} \lambda_1 &= 13.67638465, \\ \lambda_2 &= 0.04253755006, \\ \lambda_3 &= -1.718922198 \end{aligned}$$

Thus, if we solve for $X = (x_1, x_2, x_3)^t$

$$\begin{bmatrix} \lambda_i - 1 & -2 & -3 \\ -2 & \lambda_i - 4 & -7 \\ -3 & -7 & \lambda_i - 7 \end{bmatrix} \begin{bmatrix} x_1 \\ x_2 \\ x_3 \end{bmatrix} = 0, \qquad i = 1, 2, 3, \tag{4.25}$$

we obtain three eigenvectors v_i *corresponding to distinct eigenvalues* λ_i*:*

$$\begin{aligned} v_1 &= [0.2726492903, 0.6002754954, 0.7518854280]^t, \\ v_2 &= [-0.9454336145, 0.3120486268, 0.9370664388 \times 10^{-1}]^t, \\ v_3 &= [0.1783750127, 0.7364068085, -0.6526004676]^t. \end{aligned}$$

Thus, the transition matrix P *is the inverse of the matrix* $\mathbf{B}$ *whose columns are these eigenvectors:*

$$\begin{aligned} P &= \begin{bmatrix} 0.2726492903 & -0.9454336145 & 0.1783750135 \\ 0.6002754954 & 0.3120486268 & 0.7364068085 \\ 0.7518854280 & 0.9370664388 \times 10^{-1} & -0.6526004676 \end{bmatrix}^{-1} \\ &= \begin{bmatrix} 0.2726492896 & 0.6002754932 & 0.7518854264 \\ -0.9454336146 & 0.3120486268 & 0.9370664382 \times 10^{-1} \\ 0.1783750127 & 0.7364068056 & -0.6526004659 \end{bmatrix}. \end{aligned}$$

Thus, we see that $P^{-1} = P^t$*, i.e.,* P *is an orthogonal matrix. We also easily check that:*

$$PAP^t = \begin{bmatrix} 13.67638465 & & \\ & 0.04253755006 & \\ & & -1.718922198 \end{bmatrix}, \tag{4.26}$$

is the diagonal matrix whose diagonal entries consist of those eigenvalues.

Here is an immediate useful computational application of diagonalization of a square matrix A:

COROLLARY 4.7. *If* $D = PAP^{-1}$ *is diagonal, then*

$$A^k = P^{-1} D^k P.$$

□

EXAMPLE 4.17. *If A is the matrix in the above example, then*

$$\begin{aligned} A^k &= \begin{bmatrix} 1 & 2 \\ 1 & -1 \end{bmatrix} \begin{bmatrix} 5 & 0 \\ 0 & 2 \end{bmatrix}^k \begin{bmatrix} \frac{1}{3} & \frac{2}{3} \\ \frac{1}{3} & -\frac{1}{3} \end{bmatrix} \\ &= \begin{bmatrix} 1 & 2 \\ 1 & -1 \end{bmatrix} \begin{bmatrix} 5^k & 0 \\ 0 & 2^k \end{bmatrix} \begin{bmatrix} \frac{1}{3} & \frac{2}{3} \\ \frac{1}{3} & -\frac{1}{3} \end{bmatrix} \\ &= \frac{1}{3} \begin{bmatrix} 5^k + 2 \cdot 2^k & 2(5^k - 2^k) \\ 5^k - 2^k & 2 \cdot 5^k + 2^k \end{bmatrix}. \end{aligned}$$

Exercises for Section 4.6

EXERCISE 4.23.

(1) *Find the eigenvectors and eigenvalues of the matrix*

$$A = \begin{bmatrix} 2 & 1 \\ 1 & 2 \end{bmatrix}.$$

(2) *Find a matrix P such that PAP^{-1} is diagonal.*
(3) *Compute A^{30}.*

EXERCISE 4.24. *Diagonalize the rotation matrix $\begin{bmatrix} \cos\theta & -\sin\theta \\ \sin\theta & \cos\theta \end{bmatrix}$ using complex numbers.*

EXERCISE 4.25. *Prove that if A, B are square matrices and if A is invertible then AB is similar to BA.*

EXERCISE 4.26. *In each case, if the matrix is diagonalizable, find a matrix P such that PAP^{-1} is diagonal:*

(1) $\begin{bmatrix} -2 & 2 \\ -2 & 3 \end{bmatrix}$,

(2) $\begin{bmatrix} 1 & i \\ -i & 1 \end{bmatrix}$ *where i is the square root of -1,*

(3) $\begin{bmatrix} 1 & 2 & 3 \\ & 4 & 5 \\ & & 6 \end{bmatrix}$,

(4) $\begin{bmatrix} 0 & 0 & 1 \\ 1 & 0 & 0 \\ 0 & 1 & 0 \end{bmatrix}$.

EXERCISE 4.27. *Prove that a linear operator T is nilpotent if and only if there is a basis of V such that the matrix of T is upper triangular, with diagonal entries being zero. A matrix A is called* ***nilpotent*** *if there exists $n > 0$ such that $A^n = 0$.*

EXERCISE 4.28. *Derive the formula for a linear recurrence of order 2: $F_0 = x_0,\ F_1 = x_1,\ F_n := aF_{n-1} + bF_{n-2}$ for $n \geq 2$, where $a \neq 0,\ b \neq 0$, and $b \neq -a^2/4$. [Hint: Consider the linear operator $A = \begin{bmatrix} a & b \\ 1 & 0 \end{bmatrix} : \mathbf{R}^2 \longrightarrow \mathbf{R}^2$ and prove by mathematical induction on n that for all $n \in \mathbf{N}$, $\begin{bmatrix} F_{n+1} \\ F_n \end{bmatrix} = A^n \begin{bmatrix} x_1 \\ x_0 \end{bmatrix}$. Thus, if we can "expand" A^n into, say, $\begin{bmatrix} * & * \\ g & h \end{bmatrix}$, then $F_n = gx_1 + hx_2$ is the desired formula.]*

CHAPTER 5

Bilinear Forms

In our study of orthogonal matrices in Section 6 of Chapter 3, we have officially introduced dot product of column vectors in our standard or reference vector space $\mathbb{R}^n$ with respect to the standard basis $(e_1, ..., e_n)$, and it allowed us to attach a richer geometric structure to the real n-space $\mathbb{R}^n$. We now note that the dot product is in fact a map: $\mathbb{R}^n \times \mathbb{R}^n \longrightarrow \mathbb{R}$ satisfying certain defining properties. Since any finite-dimensional vector space V is isomorphic to $\mathbb{R}^n$ for some positive integer n, we want to extend the concept of dot product to a general finite-dimensional vector space V over a general field F. We will examine and separate those underlying concepts behind Proposition 4.11 very diligently, and naturally extend them to a vector space over the field $\mathbb{C}$ of complex numbers, so that, for instance, the *positivity axiom* can be imposed.

1. Definition of Bilinear Form

We start our discussion with a

DEFINITION 5.1. *Let V be a vector space over a field F. A **bilinear form** on V is a function of two variables on V, with values in the field*

$$f : V \times V \longrightarrow F,$$

satisfying the following bilinear axioms:

(Linearity in the first variable)

$$(5.1)\quad f(v_1 + v_2, w) = f(v_1, w) + f(v_2, w) \quad \textit{and} \quad f(cv, w) = cf(v, w),$$

(Linearity in the second variable)

$$(5.2)\quad f(v, w_1 + w_2) = f(v, w_1) + f(v, w_2) \quad \textit{and} \quad f(v, cw) = cf(v, w),$$

for each $v, w, v_i, w_i \in V$ *and each* $c \in F$.

NOTATION 5.1. *We will frequently use the notation*

$$\langle u, v \rangle$$

to designate the value $f(u, v)$ *of the form* f.

DEFINITION 5.2. *A form* $\langle\ ,\ \rangle$ *is called* ***symmetric*** *if*

$$\langle v, w \rangle = \langle w, v \rangle, \tag{5.3}$$

and ***skew-symmetric*** *if*

$$\langle v, v \rangle = 0, \tag{5.4}$$

for each $v, w \in V$.

EXAMPLE 5.1. *The dot product* $X \cdot Y$ *of two column vectors* $X, Y \in \mathbb{R}^n$ *is certainly a bilinear form on* $\mathbb{R}^n$, *which is obviously symmetric. In addition to this, it satisfies* $X \cdot X = 0$ *if* $X = 0$ *and the* ***positivity axiom:***

$$X \cdot X > 0 \qquad \text{if } X \neq 0$$

But since which element is bigger than others is impossible to think of even when F *is the field of complex numbers, there is no way of safely generalizing the positivity axiom unless the field* F *of scalars is the set* $\mathbb{R}$ *of real numbers. That is, we can think of positivity axiom only when our vector space is a* real *vector space, first of all.*

LEMMA 5.1. *If a form is skew-symmetric, then*

$$\langle v, w \rangle = -\langle w, v \rangle. \tag{5.5}$$

for each $v, w \in V$

PROOF. Just expand

$$0 = \langle v + w, v + w \rangle.$$

□

REMARK 5.1. *If the characteristic of F is not 2, i.e., 2 is the smallest number such that $2 \cdot 1 = 0$, then (5.4) and (5.5) are equivalent. If the characteristic of F is 2, then (5.5) and (5.3) are the same but most symmetric forms do not satisfy (5.4).*

If the form is symmetric, then linearity in the second variable follows from linearity in the first variable.

EXAMPLE 5.2. *The main examples of bilinear forms are the forms on the space F^n of column vectors obtained as follows: Let A be an $n \times n$ matrix with entries in F. Define a* **product** *of column vectors:*

$$\langle X, Y\rangle = X^t AY\colon F^n \times F^n \longrightarrow F. \tag{5.6}$$

Here it is assumed that F^n is equipped with the standard basis $(e_1, ..., e_n)$, and our dot product is one of such products with $A = I$ and $F = \mathbb{R}$.

PROPOSITION 5.1. *The form (5.6) is symmetric if and only if the matrix $A = (a_{ij})$ is symmetric.*

PROOF. Assume that A is symmetric. Then since formX^tAY is a 1×1 matrix, it is equal to its transpose; hence

$$\langle X, Y\rangle = \langle X, Y\rangle^t = Y^t A^t X = Y^t AX = \langle Y, X\rangle.$$

On the other hand, if the form is symmetric, then in particular with $X = e_i, Y = e_j$,

$$a_{ij} = e_i^t Ae_j = \langle e_i, e_j\rangle = \langle e_j, e_i\rangle = e_j^t Ae_i = a_{ji},$$

for each i, j. Thus, A is symmetric. □

EXAMPLE 5.3. *Thus the dot product is symmetric because its associated matrix, the identity matrix I, is symmetric. Moreover, I is invertible.*

Let $\langle\ ,\ \rangle$ be a bilinear form on a vector space V over a field F and let $\mathbf{B} = (v_1, ..., v_n)$ be its basis. Then we can associate to this basis $\mathbf{B}$ a square matrix $A = (a_{ij})$ such that the associated bilinear form XA^tY in (5.6) of the matrix A on the coordinate vectors X, Y of any two vectors

v, w of V with the isomorphism $v \mapsto X$, where $v = \mathbf{B}X, w = \mathbf{B}Y$, satisfies

$$\langle\, v, w \,\rangle = X^t AY = \langle\, X, Y \,\rangle. \tag{5.7}$$

For this, we define for each i, j

$$a_{ij} = \langle\, v_i, v_j \,\rangle. \tag{5.8}$$

Then

$$\begin{aligned} \langle\, v, w \,\rangle &= \langle\, \mathbf{B}X, \mathbf{B}Y \,\rangle = \langle\, \sum_i v_i x_i, \sum_j v_j y_j \,\rangle \\ &= \sum_{i,j} x_i \langle\, v_i, v_j \,\rangle y_j = \sum_{i,j} x_i a_{ij} y_j = X^t AY \\ &= \langle\, X, Y \,\rangle. \end{aligned}$$

Thus, if we identify F^n with V using a given basis $\mathbf{B}$ as in Proposition 3.14 of Chapter 3, the bilinear form $\langle\ ,\ \rangle$ corresponds to $X^t AY$.

DEFINITION 5.3. *We say that $X^t AY$ or A **represents** the bilinear form $\langle\ ,\ \rangle$ on a vector space with respect to a basis $\mathbf{B}$, and we call A the **matrix of the form** with respect to $\mathbf{B}$, and call $X^t AX$ its associated **quadratic form** on F^n with respect to $\mathbf{B}$.*

REMARK 5.2. *We also note that*

a: *A different basis $\mathbf{B}$ gives rise to a different matrix A.*
b: *The form $\langle\ ,\ \rangle$ is symmetric if and only if A is symmetric.*
c: *If the matrix A of the form is symmetric with respect to some basis, then its matrix with respect to any basis is symmetric.*

The last statement holds, since the symmetry of a bilinear form does not depend on a particular basis.

Since a different basis gives rise to a different matrix A, we first want to know exactly how these matrices, although they represent the same bilinear form with respect to different bases, are related to each other under change of basis:

Let A be the matrix associated to an old basis $\mathbf{B}$ and A' associated to a new basis $\mathbf{B}'$. Thus, we have

$$\langle\ v, w\ \rangle = X^tAY = X'^tA'Y',$$

for any vectors v and w in V. Here, we have several relations: $v = \mathbf{B}X = \mathbf{B}'X'$ and $w = \mathbf{B}Y = \mathbf{B}'Y', X' = PX, Y' = PY$, and $\mathbf{B} = \mathbf{B}'P$. Using some of these relations, if we express, say, X and Y in terms of X' and Y' in F^n with the standard basis $\mathbf{E} = (e_1, ..., e_n)$, then by choosing e_i for X' and e_j for Y', for $i, j = 1, 2, ..., n$, we can express A in terms of A'.

$$X^tAY = (P^{-1}X')^tA(P^{-1}Y') = X'^t(P^{-1})^tA'P^{-1}Y'.$$

Thus we have expressed the equality $X^tAY = X'^tA'Y'$ in terms of X' and Y':

$$X'^t(P^{-1})^tAP^{-1}Y' = X'^tA'Y' \qquad \text{for each } X', Y'.$$

Therefore, if we choose e_i for X' and e_j for Y' for $i, j = 1, 2, ..., n$, we explicitly see that

$$(P^{-1})^tAP^{-1} = A'. \tag{5.9}$$

With $Q = (P^{-1})^t$, this establishes and proves:

PROPOSITION 5.2. *Let A be the matrix of a bilinear form with respect to a basis. Then the matrix which represents the same form with respect to different bases are the matrices $A' = QAQ^t$, where Q is an arbitrary invertible matrix in $GL_n(F)$.* □

DEFINITION 5.4. *We say that a square matrix A is* ***congruent*** *to B if there exists an invertible matrix $Q \in GL_n(F)$ such that $B = QAQ^t$.*

We recall that if B is additionally diagonal, then A is called **congruently diagonalizable** (Definition 4.26). Thus, two orthogonally similar matrices are congruent, but not conversely in general.

In particular, we apply this proposition to the case where $F = \mathbb{R}$ and our bilinear form is the dot product $X \cdot Y$ of columns vectors $X, Y \in \mathbb{R}^n$ (with respect to the standard basis $\mathbf{E}$) In this case, since the matrix associated to the dot product is the identity matrix I, I is changed to $A' = (P^{-1})^t I P^{-1}$ when we change our old basis, the standard basis $\mathbf{E} = (e_1, ..., e_n)$, to a new basis $\mathbf{B} = P^{-1}$ for some invertible matrix P: $\mathbf{E} = \mathbf{B}P$ and $X' = PX$.

COROLLARY 5.1. *Let P be a transition matrix of change of basis from the standard $\mathbf{E}$ to a basis $\mathbf{B}'$. Then $X \cdot Y = PX \cdot PY$, i.e., dot product is preserved by left multiplication by P, if and only if P is an orthogonal matrix.*

PROOF. We note that the dot product $X \cdot Y$ of column vectors X and Y *represents* a bilinear form f on $\mathbf{R}^n$ with respect to the standard basis $\mathbf{E}$. Since the matrix I of the dot product $X \cdot Y$ is changed to $A' = (P^{-1})^t I P^{-1}$ under the change of basis $X' = PX$, another dot product $X' \cdot Y'$ represents the same form f with respect to a new basis $\mathbf{B}$, that is to say that $PX \cdot PY = X \cdot Y$, if and only if $A' = I$ if and only if P is orthogonal. □

REMARK 5.3. *We note that we have in fact proved the equivalence of part a and part b of Proposition 4.11.*

If the transition matrix P is not orthogonal, it is quite confusing:

EXAMPLE 5.4. *Consider $\mathbb{R}^2$ with a new basis $\mathbf{B}' = ((1,1)^t, (0,1)^t)$. Then*

$$P^{-1} = \begin{bmatrix} 1 & 0 \\ 1 & 1 \end{bmatrix}; \qquad P = \begin{bmatrix} 1 & 0 \\ -1 & 1 \end{bmatrix}.$$

Hence P is not orthogonal, but

$$A' = (P^{-1})^t I P^{-1} = \begin{bmatrix} 1 & 1 \\ 0 & 1 \end{bmatrix} \begin{bmatrix} 1 & 0 \\ 1 & 1 \end{bmatrix} = \begin{bmatrix} 2 & 1 \\ 1 & 1 \end{bmatrix}$$

still represents *a bilinear form, called* the *dot product on $\mathbb{R}^2$ (i.e., with respect to the standard basis), with respect to a new basis $\mathbf{B}'$. That is, $X \cdot Y = X'^t A' Y'$ holds, where $X' = PX$.*

Now, suppose that we are given a bilinear form $\langle\ ,\ \rangle$ on a *real* vector space V. We would like to know under what condition this form is represented by a dot product with respect to a certain basis. For this, we start with an arbitrary basis $\mathbf{B}$ so that the form is represented as X^tAY, where any vector $v \in V$ has its coordinate vector X with respect to $\mathbf{B}$, $v = \mathbf{B}X$. Thus, the problem is how to change this basis to another basis $\mathbf{B}'$ in such a way that the new matrix for the same form with respect to $\mathbf{B}'$ is the identity matrix, if that is possible. More precisely, we have: $\langle\ v, w\ \rangle = X^tAY$ with respect to $\mathbf{B}$, $\langle\ v, w\ \rangle = X'^tA'Y'$ with respect to $\mathbf{B}'$, and $\mathbf{B} = \mathbf{B}'P$; hence $X' = PX$ and $Y' = PY$. Thus,

$$\langle\ v, w\ \rangle = X^tAY = (P^{-1}X')^tA(P^{-1}Y') = X'^t(P^{-1})^tAP^{-1}X'.$$

Under these usual settings, we have:

COROLLARY 5.2. *$\langle\ v, w\ \rangle = X' \cdot Y'$ if and only if the equation $I = (P^{-1})^tAP^{-1}$ in P has a solution for P.* □

Or equivalently,

COROLLARY 5.3. *The matrices A of a bilinear form on a real vector space V such that $X^tAY = X' \cdot Y'$ are the matrices $A = P^tP$, where P is invertible.* □

REMARK 5.4. *But the immediate problem with this corollary is that it is not clear how we are going to figure out those invertible matrices P such that $A = P^tP$. Thus, we need to develop a method which guarantees how to find at least one such invertible matrix P, called* ***Gram-Schmidt orthogonalization process.***

Since we want to represent our real bilinear form with a dot product, our form $\langle\ v, w\ \rangle = X^tAY$ must be necessarily symmetric and positive definite. Thus, the matrix A of the form must be symmetric.

DEFINITION 5.5. *A real matrix A is* ***positive definite*** *if its associated* quadratic form *is positive definite, i.e.,*

$$X^tAX > 0 \qquad \text{if } X \neq 0.$$

A bilinear form $\langle\ ,\ \rangle$ *on a finite-dimensional real vector space* V *is called* ***positive definite*** *if*

$$\langle\ v, v\ \rangle > 0$$

for each non-zero vector $v \in V$.

Thus, a real matrix A is positive definite if and only if the form $\langle X, Y\ \rangle = X^t AY$ it defines on $\mathbb{R}^n$ is positive definite. Also, the form $\langle\ ,\ \rangle$ is positive definite if and only if its matrix A with respect to *any* basis is a positive definite matrix.

DEFINITION 5.6. *A finite-dimensional* real *vector space is called an* ***inner product space*** *or* ***Euclidean space*** *if it has a positive definite and symmetric bilinear form on it.*

DEFINITION 5.7. *Two vectors* v, w *are called* ***orthogonal*** *with respect to a symmetric bilinear form if* $\langle\ v, w\ \rangle = 0$.

NOTATION 5.2. *Orthogonality of two vectors is often denoted as*

$$v \perp w.$$

This definition extends the concept of orthogonality which we have already seen when the form is dot product on $\mathbb{R}^n$.

DEFINITION 5.8. *A basis* $\mathbf{B} = (v_1, ..., v_n)$ *of* V *is called an* ***orthonormal basis*** *with respect to the form if*

$$\langle\ v_i, v_j\ \rangle = 0 \text{ if } i \neq j, \qquad \text{and } \langle\ v_i, v_i\ \rangle = 1 \text{ for each } i.$$

Then, a basis $\mathbf{B}$ is orthonormal if and only if the matrix of the form with respect to $\mathbf{B}$ is the identity matrix.

For a real matrix, being symmetric and being positive definite are two different properties.

COROLLARY 5.4. *A bilinear form* $\langle\ ,\ \rangle$ *on a real finite-dimensional vector space* V *is symmetric if and only if it is diagonalizable.*

PROOF. By Theorem 4.3 in Chapter 4, every real symmetric matrix is orthogonally diagonalizable; thus we already know the $\Rightarrow$ direction holds. Since a diagonal matrix is symmetric the converse is obvious. $\square$

COROLLARY 5.5. *Let* $\langle \ , \ \rangle$ *be a* positive definite *bilinear form on a* real *finite-dimensional vector space* V. *If it is moreover symmetric, then it's associated matrix* A *(with respect to any basis) is invertible.*

PROOF. By Theorem 4.3 in Chapter 4, A is orthogonally diagonalizable to a diagonal matrix D. Since the form is positive definite, each of the diagonal entries of D must be non-zero, and hence A is invertible. $\square$

REMARK 5.5. *This also proves that part c* $\Rightarrow$ *part b of Theorem 5.2 below, but in contrast to Theorem 5.1 it does not show how to obtain the corresponding orthonormal basis of* V *explicitly.*

REMARK 5.6. *We recall Theorem 4.3 in Chapter 4: Every* real symmetric $n \times n$ *matrix is congruently diagonalizable. If we add one more condition that it is also positive definite, then either by the above corollary or by the following Theorem 5.1, it is congruently diagonalizable to the identity matrix, though the involved invertible matrix* P *may not be orthogonal.*

Now we are ready to state and prove the so called **Gram-Schmidt procedure** for constructing an orthonormal basis, starting with an arbitrary basis $\mathbf{B} = (v_1, ..., v_n)$.

THEOREM 5.1. *Let* $\langle \ , \ \rangle$ *be a* positive definite symmetric *form on a finite-dimensional* real *vector space* V. *Then there exists an orthonormal basis for* V.

PROOF. Let $\mathbf{B} = (v_1, ..., v_n)$ be a basis of V. Assume that $\mathbf{B}$ is not orthonormal. Since the form is positive definite, we first normalize v_1 so that $\langle v_1, v_1 \rangle = 1$: That is, we want $\langle cv_1, cv_1 \rangle = c^2 \langle v_1, v_1 \rangle = 1$, so we choose $c_1 = \langle v_1, v_1 \rangle^{-1/2}$. Thus, $w_1 = c_1 v_1$ has length 1 and thus we replace v_1 with w_1 in $\mathbf{B} = (v_1, ..., v_n)$ to get an intermediate basis $\mathbf{B}_1 =$

$(w_1, v_2, ..., v_n)$. Then we define $w_2' = v_2 - \langle\ v_2, w_1\ \rangle w_1$. Then $\langle\ w_2', w_1\ \rangle = 0$. Thus we normalize w_2' to a new vector w_2 of length 1, and we replace v_2 in $\mathbf{B}_1 = (w_1, v_2, ..., v_n)$ with w_2 to get a basis $\mathbf{B}_2 = (w_1, w_2, v_3, ..., v_n)$. We note that $\mathbf{B}_2$ is indeed a basis, since $c_1w_1 + c_2w_2 + c_3v_3 + ... + c_nv_n = 0$ implies

$$(c_1 - c_2\frac{\langle\ v_2, w_1\ \rangle}{\langle\ w_2', w_2'\ \rangle^{1/2}})w_1 + c_2\frac{1}{\langle\ w_2', w_2'\ \rangle^{1/2}}v_2 + ... + c_nv_n = 0,$$

hence every c_i are zero since $\mathbf{B}_1 = (w_1, v_2, ..., v_n)$ is a basis of V.

This shows how we must proceed: Suppose that the first $k-1$ vectors $w_1, ..., w_{k-1}$ are orthonormal and that $\mathbf{B}_{k-1} = (w_1, ..., w_{k-1}; v_k, ...v_n)$ is a basis of V. Then we define

$$w_k' = v_k - \langle\ v_k, w_1\ \rangle w_1 - \langle\ v_k, w_2\ \rangle w_2 - ... - \langle\ v_k, w_{k-1}\ \rangle w_{k-1}.$$

Then normalize w_k' to get w_k; then replace v_k in $\mathbf{B}_{k-1}$ with w_k to get a new basis $\mathbf{B}_k = (w_1, ..., w_k; v_{k+1}, ...v_n)$, where the first k vectors are orthogonal. We then continue this procedure until we get $\mathbf{B}_n = (w_1, ..., w_n)$, this basis is the orthogonal basis we claimed. □

THEOREM 5.2. *Let A be a* real *$n \times n$ matrix. Then the following statements are equivalent to each other:*

- **a:** *A represents dot product, with respect to some basis $\mathbf{B}'$ of $\mathbb{R}^n$, i.e., its form $X^tAY = X' \cdot Y'$*
- **b:** *There is an invertible matrix $P \in GL_n(\mathbb{R})$ such that $A = P^tP$.*
- **c:** *A is symmetric and positive definite.*

PROOF. We have shown the equivalence of part a and part b in Corollary 5.3. We also have seen that part c is a necessary condition for part a. To prove part c implies part a, we apply Theorem 5.1 as follows: Since A is symmetric and positive definite, so is the form $\langle\ X, Y\ \rangle = X^tAY$ on $\mathbb{R}^n$ with a non-orthonormal basis $\mathbf{B}$; hence there exists an orthonormal basis $\mathbf{B}'$ of $\mathbb{R}^n$ with respect to this form $\langle\ X, Y\ \rangle = X^tAY$ (i.e., not with respect to the usual dot product on $\mathbb{R}^n$). Then by (5.9), the matrix A' of the form $\langle\ X, Y\ \rangle$ with respect to this new basis $\mathbf{B}'$ satisfies $P^tA'P = A$. But since

$\mathbf{B}'$ is orthonormal, its matrix A' is the identity matrix I by definition. Thus, part c implies part b, and hence part a, because we already know that part b and part a are equivalent. □

COROLLARY 5.6. *A real symmetric positive definite matrix A is invertible.* □

REMARK 5.7. *The proof of the following theorem shows explicitly how to construct P satisfying $A = P^tP$ in Theorem 5.2.*

THEOREM 5.3. *Let A be a* real *$n \times n$ matrix. If it is moreover* symmetric, *then the following statements are equivalent to each other:*

a: *A is positive definite.*
b: *Eigenvalues of A are all positive.*
c: *The determinant of each principal submatrix A_k of A is positive.*
d: *The diagonal entries of the diagonal matrix D in a decomposition $A = LDL^t$ of A by a lower triangular matrix L (without row exchange) are all positive.*
e: *There exists an invertible matrix P such that $A = P^tP$.*

PROOF. We first note that by Theorem 4.3, there exists an orthogonal matrix P such that $D = P^tAP$ is a diagonal matrix with diagonal entries λ_i. Thus,

$$X^tAX = X^t(P^{-1})^tDP^{-1}X = Y^tDY = \sum_i \lambda_i y_i^2,$$

where $Y = P^{-1}X = (y_1, ..., y_n)^t$.

Part a $\Leftrightarrow$ part b: Thus, for the $\Longrightarrow$ direction, we choose $Y = e_i$ for each $i = 1, 2, ..., n$. The $\Longleftarrow$ direction is obvious.

Part b $\Rightarrow$ part c: If real eigenvalues of A are all positive, then $\det A =$ there product > 0. For each $k = 1, 2, ..., n$, consider all the vectors of the form: $X = (x_1, ..., x_k, 0, ..., 0)^t$, where $X_k = (x_1, ..., x_k)^t$ is any vector in $\mathbb{R}^k$. Then by block multiplication of matrices, we have

$$X^tAX = [X_k^t, 0]\begin{bmatrix} A_k & 0 \\ 0 & 0 \end{bmatrix}\begin{bmatrix} X_k \\ 0 \end{bmatrix} = X_k^tA_kX_k.$$

Thus, A is positive definite if and only if each A_k is positive definite for $k = 1, ..., n$. Thus, via part b, all the eigenvalues of A_k are positive; hence all the $\det A_k$ are positive.

Part c $\Rightarrow$ part d: By hypothesis each A_k is invertible. Thus A is invertible. Applying only elementary operations E_i of type II, we get

$$E_k \cdots E_1 A = U_1$$

where each diagonal entry of the upper triangular matrix U_1 is non-zero. Thus we get,

$$A = LU_1 = LDU_1,$$

and moreover, this LDU decomposition is unique. Then since A is symmetric, we get

$$A = LDL^t,$$

where L is a lower triangular invertible matrix with 1's in the diagonal entries and D is a diagonal matrix with the pivots d_k on the diagonal entries. Then $\det(L) = 1$ and by hypothesis $\det(A) = \det(D) > 0$ and $d_1 = \det A_1 > 0$, and hence for each $k = 2, ..., n$, the kth diagonal entry d_k of D is simply the ratio

$$d_k = \frac{\det A_k}{\det A_{k-1}}.$$

Thus all the diagonal entries d_k of D are positive.

Part d $\Rightarrow$ part e. Let D be a diagonal matrix with each diagonal entry $d_i > 0$. Define D' another diagonal entries with each diagonal entry $\sqrt{d_i} > 0$. Then $D = D'^2$, and hence $A = LDL^t = (LD')D'^tL^t = (LD')(LD')^t$, where $P = LD'$ is obviously an invertible matrix.

Part e $\Rightarrow$ part a. For any $X \neq 0$,

$$X^tAX = X^tP^tPX = (PX)^t(PX) = (PX) \cdot (PX) > 0.$$

□

EXAMPLE 5.5. *(MAPLE) The following lines compute the orthonormal basis from a given basis, using the inner product:*

—————————

interface(prettyprint=0);
with(linalg):
v1:=vector([-1,2,3,0]);
v2:=vector([0,1,2,1]);
v3:=vector([2,-1,-1,1]);
 # The following line verifies whether these vectors are a basis or not.
basis({v1,v2,v3});
 # Once it is verified, we apply the following command to compute the orthogonal basis:
result:=GramSchmidt([v1,v2,v3],normalized);

————————

Exercises for Section 5.1

EXERCISE 5.1. *Let A and B be real $n \times n$ matrices. Prove that if $X^tAY = X^tBY$ for each vectors X, Y in $\mathbb{R}^n$, then $A = B$.*

EXERCISE 5.2. *Apply the Gram-Schmidt process to the basis*

$$(1,1,0)^t, (1,0,1)^t, (0,1,1)^t,$$

when the form is dot product.

EXERCISE 5.3. *Let $A = \begin{bmatrix} 2 & 1 \\ 1 & 2 \end{bmatrix}$. Find an orthonormal basis for $\mathbb{R}^2$ with respect to the form X^tAY.*

EXERCISE 5.4. *Prove that every real square matrix is the sum of a symmetric matrix and a skew-symmetric matrix in exactly one way.*

EXERCISE 5.5. *Let $\langle\ ,\ \rangle$ be a bilinear form on a real vector space V. Show that there exists a symmetric form $(\ ,\ )$ and a skew-symmetric form $[\ ,\ ]$ such that $\langle\ ,\ \rangle = (\ ,\ ) + [\ ,\ ]$.*

EXERCISE 5.6. *Let $\langle\ ,\ \rangle$ be a symmetric bilinear form on a real vector space V over a field F. The function $q\colon V \longrightarrow \mathbb{R}$ defined by $q(v) = \langle v, v \rangle$ is called the* ***quadratic form****. associated to the bilinear form. Show how to recover the bilinear form from q by expanding $q(v + w)$.*

EXERCISE 5.7. *Let* $A = \begin{bmatrix} 4 & 2 & 2 \\ 2 & 4 & 2 \\ 2 & 2 & 4 \end{bmatrix}$. *Compute* P *such that* $A = P^tP$.

2. Symmetric Forms: Orthogonality

In contrast to the previous section, we consider a finite-dimensional real vector space V on which a symmetric bilinear form $\langle\ ,\ \rangle$ is given, but we drop the assumption that it is positive definite.

DEFINITION 5.9. *A symmetric bilinear form* $\langle\ ,\ \rangle$ *on a real finite-dimensional vector space* V *is called* ***indefinite*** *if* $\langle\, v, v\,\rangle$ *is either positive, zero, or negative for each non-zero vector* $v \in V$.

EXAMPLE 5.6. *The Lorentz form*

$$X^tAY = x_1y_1 + x_2y_2 + x_3y_3 - c^2x_4y_4$$

of physics is a typical example of an indefinite form on "space-time" $\mathbb{R}^4$. *The coefficient c representing the speed of light can be normalized to* 1, *and then the matrix of the form with respect to the given basis becomes*

$$\begin{bmatrix} 1 & & & \\ & 1 & & \\ & & 1 & \\ & & & -1 \end{bmatrix}. \tag{5.10}$$

PROPOSITION 5.3. *Suppose the real symmetric form* $\langle\ ,\ \rangle$ *is not identically zero. Then there exists a vector* $v \in V$ *that is not self-orthogonal:* $\langle\, v, v\,\rangle \neq 0$.

PROOF. Since it is not identically zero, there exist v, w such that $\langle\, v, w\,\rangle \neq 0$. If $\langle\, v, v\,\rangle \neq 0$ or $\langle\, w, w\,\rangle \neq 0$, then we are done. Now suppose that $0 = \langle\, v, v\,\rangle = \langle\, w, w\,\rangle$. Then their sum $v + v$ is not self-orthogonal:

$$\langle\, v + w, v + w\,\rangle = \langle\, v, v\,\rangle + \langle\, v, w\,\rangle + \langle\, w, v\,\rangle + \langle\, w, w\,\rangle = 2\langle\, v, w\,\rangle \neq 0.$$

□

DEFINITION 5.10. *Let V be a real vector space with a symmetric bilinear form on it.*

(1) *An **orthogonal complement** of a subspace W of V is*

$$W^{\perp} = \{v \in V : \langle\ v, W\ \rangle = 0\}.$$

(2) *A **null space** of the form is*

$$N = \{v \in V : \langle\ v, V\ \rangle = 0\}.$$

(3) *An element of the null space is called a **null vector** for the given form.*

(4) *A symmetric bilinear form is called **nondegenerate** if its null space is $\{0\}$.*

Thus, the null space of V is just $V^{\perp}$.

LEMMA 5.2. *$W^{\perp}$ is a subspace of V.*

PROOF. Exercise. □

PROPOSITION 5.4. *Let $w \in V$ be a non-zero vector such that $\langle\ w, w\ \rangle \neq 0$ and let W be its span. Then $V = W \oplus W^{\perp}$.*

PROOF. According to Proposition 3.23, we need to show two things: (a) $W \cap W^{\perp} = \{0\}$, (b) $V = W + W^{\perp}$.

The first thing is clear since cw is not orthogonal to w unless the scalar $c = 0$. Now, for any non-zero vector $v \in V$ we can write $v = \alpha w + (v - \alpha w)$. If we can show that there exists a scalar α such that $v - \alpha w \in W^{\perp}$ then we are done. For this we simply choose

$$\alpha = \frac{\langle\ v, w\ \rangle}{\langle\ w, w\ \rangle}$$

for the given $v \in V$. Then for any $\beta w \in W$, we clearly have $\langle\ v - \alpha w, \beta w\ \rangle = 0$. □

PROPOSITION 5.5. *Let A be the matrix of a symmetric form with respect to a basis. Then*

a: *The null space of the form is the set of those vectors whose coordinate vectors constitute the solution space of* $AX = 0$.

b: *The form is nondegenerate if and only if the matrix* A *is nonsingular, i.e., invertible.*

c: *If the form is positive definite, then it is nondegenerate.*

PROOF. The corresponding form via the basis is the product X^tAY. If Y is a vector such that $AY = 0$, then $X^tAY = 0$ for each X; hence Y is in the null space. Conversely, suppose that $AY \neq 0$. Then AY has at least one non-zero coordinate and e_i^tAY is the ith-coordinate of AY. Thus, there is i such that e_i^tAY is not zero. This means that Y is not in the null space, which proves part a. Part b of the proposition follows from part a. Part c is obvious. □

Here is a generalized version of Proposition 5.4:

PROPOSITION 5.6. *Let* W *be a non-zero subspace of* V, *and consider the restriction of a symmetric form* $\langle\ ,\ \rangle$ *on* V *to* W. *Suppose that this form is nondegenerate on* W. *Then* $V = W \oplus W^{\perp}$.

PROOF. If $w \in W \cap W^{\perp}$ then $\langle\, w, z\,\rangle = 0$ for each $z \in W$. Thus w is a null vector of W. But the only null vector of W is 0, since the form on W is nondegenerate. Thus, $w = 0$. This shows that $W \cap W^{\perp} = \{0\}$.

Now, we need to prove that $V = W + W^{\perp}$. Let $\mathbf{B}' = (w_1', ..., w_m')$ be a basis of W. Then by hypothesis, the restriction to W of the form on V is also symmetric and thus by Theorem 5.4 below or by Theorem 4.3 in Chapter 4, the matrix associated to $\mathbf{B}'$ is diagonalizable to a diagonal matrix $D = (d_{ij})$. Then since the restricted form on W is nondegenerate by hypothesis, D must be invertible, i.e., none of the diagonal entries of D are zero, by part b of Proposition 5.5. (Or alternatively, the restriction to W of the form on V is symmetric and nondegenerate by hypothesis, and hence by part a of Theorem 5.4 below, the diagonal matrix D associated to the restricted form on W is invertible.) Therefore, if we write $\mathbf{B} = (w_1, ..., w_m)$ for a new basis of W of which associated matrix is D, then $\langle\, w_i, w_i\,\rangle = d_{ii} \neq 0$ for each $i = 1, ..., m$ and $\langle\, w_i, w_j\,\rangle = d_{ij} = 0$ for $i \neq j$.

Thus, given $v \in V$ we define a vector w in W:

$$w := \sum_{i=1}^{m} \frac{\langle\, v, w_i \,\rangle}{\langle\, w_i, w_i \,\rangle} w_i.$$

Then we can easily check that for each $j = 1, ..., m$,

$$\langle\, v - w, w_j \,\rangle = 0,$$

that is, $v - w \in W^{\perp}$. Thus, we have established a sum

$$v = w + (v - w),$$

with $w \in W$ and $v - w \in W^{\perp}$. □

REMARK 5.8. *In the above proof we provided an alternative proof applying the following theorem, Theorem 5.4. The point of this alternative proof is that the next theorem does not require Schur's lemma, or the spectral theorem at all.*

DEFINITION 5.11. *An* ***orthogonal basis*** $\mathbf{B} = (v_1, ..., v_n)$ *for a real vector space V, with respect to a symmetric form $\langle\ ,\ \rangle$ on it, is a basis such that $v_i \perp v_j$ for all $i \neq j$.*

We note that a basis $\mathbf{B}$ is orthogonal if and only if its associated matrix A is diagonal. Thus, if a symmetric form $\langle\ ,\ \rangle$ is nondegenerate and the basis $\mathbf{B} = (v_1, ..., v_n)$ is orthogonal, then the diagonal entries of A are non-zero, so that A is invertible (part b of Proposition 5.5). If the form $\langle\ ,\ \rangle$ is not non-degenerate, A is not going to be invertible in general:

THEOREM 5.4. *Let $\langle\ ,\ \rangle$ be a* symmetric *bilinear form on a* real *finite-dimensional vector space V. Then*

- **a:** *There exists an orthogonal basis for V. More precisely, there exists a basis $\mathbf{B} = (v_1, ..., v_n)$ such that $\langle\, v_i, v_j \,\rangle = 0$ for $i \neq j$ and such that for each i, $\langle\, v_i, v_i \,\rangle = 1, -1,$ or 0.*
- **b:** *(Matrix Form) Let A be a* real symmetric *$n \times n$ matrix. Then there exists a matrix $Q \in GL_n(\mathbb{R})$ such that QAQ^t is a diagonal matrix each of whose diagonal entries is $1, -1,$ or 0.*

PROOF. Since part b follows from part a by Proposition 5.2, we need to prove part a.

If the form is identically zero, then the associated matrix is the zero matrix, which is diagonal. Suppose that the form is not identically zero. Then by Proposition 5.3, there exists a vector v_1 such that $\langle\, v_1, v_1 \,\rangle \neq 0$. Let W be the span of v_1. Then $V = W \oplus W^{\perp}$ by Proposition 5.4; hence a basis for V is obtained by combining the basis (v_1) of W with any basis $(v_2, \ldots, v_n)$ of $W^{\perp}$ (Proposition 3.25 of Chapter 3). Then the form on V can be restricted to the subspace $W^{\perp}$, and it defines a form there. Thus, we use induction on the dimension to conclude that $W^{\perp}$ has an orthogonal basis $(v_2, \ldots, v_n)$. Then $(v_1, v_2, \ldots, v_n)$ is an orthogonal basis of V because $\langle\, v_1, v_i \,\rangle = 0$ if $i > 1$ since $v_i \in W^{\perp}$.

It remains to normalize the orthogonal basis we have just constructed. If $\langle\, v_i, v_i \,\rangle \neq 0$, then we solve $1 = (\pm\langle\, v_i, v_i \,\rangle) \cdot c^2$ for c, and change the basis vector v_i to cv_i. This changes $\langle\, v_i, v_i \,\rangle$ to ± 1, and this completes the proof. □

REMARK 5.9. *Since A is real and symmetric, by Theorem 4.3 in Chapter 4 it is orthogonally diagonalizable, or equivalently, it has an orthonormal set of n eigenvectors. Thus, first computing eigenvalues and eigenvectors for its diagonalization, we can compute the invertible matrix Q in part b of Theorem 5.4, where we will need to normalize the diagonal matrix as a final step. (Cf. Example 4.16 in Chapter 4.)*

We can permute an orthogonal basis $\mathbf{B}$ in such a way that the indices with $\langle\, v_i, v_i \,\rangle = 1$ are the first ones, $\langle\, v_i, v_i \,\rangle = -1$ are the second ones, and $\langle\, v_i, v_i \,\rangle = 0$ are the last ones. Then the matrix A of the form will be

$$A = \begin{bmatrix} I_p & & \\ & -I_m & \\ & & 0_z \end{bmatrix}, \tag{5.11}$$

where p is the number of $+1$'s, m the number of -1's, and z the number of 0's such that $p + m + z = n$. Then the next theorem tells us that these numbers are uniquely determined by the form or by the matrix A:

THEOREM 5.5. *(Sylvester's Law) The numbers, p, m, z appearing in (5.11) are uniquely determined by the form. In other words, they do not depend on the choice of orthogonal basis* **B** *such that* $\langle v_i, v_i \rangle = +1, -1,$ *or* 0.

The pair of integers (p, m) is called the *signature* of the form.

PROOF. Write $r = p + m$. (We note that r is the rank of the matrix.) Let $(v_1, ..., v_n)$ be an orthogonal basis of V of the type under consideration, i.e., so that its associated matrix is (5.11). We will first show that the number z is uniquely determined, by proving that the vectors $v_{r+1}, ..., v_n$ form a basis for the null space $N = V^{\perp}$; this will show that z is not dependent of the choice of a basis.

Let w be an arbitrary vector in $N = V^{\perp} \subset V$. Then $w = c_1v_1 + ... + c_nv_n$ for unique scalars c_i's. We note that $\langle w, v_i \rangle = 0$ for each $i = 1, ..., n$. Then since $\langle v_i, v_j \rangle = 0$ for $i \neq j$, we have $\langle w, v_i \rangle = c_i \langle v_i, v_i \rangle$. By hypothesis, $\langle v_i, v_i \rangle = 0$ if and only if $i > r$. Thus, we must have $c_i = 0$ for each $i \leq r$. On the other hand, we also know by hypothesis that for $r < i \leq n$, $\langle v_i, v_j \rangle = 0$ for each $j = 1, ..., n$. This exactly means that $v_{r+1}, ..., v_n$ are in fact elements of the null space N. This shows that $(v_{r+1}, ..., v_n)$ spans N; hence they form a basis of N since they are linearly independent.

Then the equation $p + m + z = n$ uniquely determines the sum $p + m$; and hence we need to prove one of the two remaining integers p, m is uniquely determined.

Suppose that a second such basis $(v'_1, ..., v'_n)$ is given and it gives integers p', m', and $z' = z$. Our strategy is to show that the $p + (n - p')$ vectors

$$v_1, ..., v_p; v'_{p'+1}, ..., v'_n \tag{5.12}$$

are linearly independent. Then since V has dimension n, it will follow that $p + (n - p') \leq n$; hence $p \leq p'$, and, interchanging the roles of p and p', it follows that $p = p'$.

Let

$$b_1 v_1 + ... + b_p v_p = c_{p'+1} v_{p'+1} + ... + c_n v'_n \tag{5.13}$$

be a linear relation between the vectors in (5.12). Let v denote the vector defined by either of these two expressions. We compute $\langle\, v, v \,\rangle$ in two different ways. The left-hand side gives

$$\langle\, v, v \,\rangle = b_1^2 \langle\, v_1, v_1 \,\rangle + ... b_p^2 \langle\, v_p, v_p \,\rangle = b_1^2 + ... + b_p^2 \geq 0,$$

while the right-hand side gives

$$\langle\, v, v \,\rangle = c_{p'+1}^2 \langle\, v_{p'+1}, v_{p'+1} \,\rangle + ... + c_n^2 \langle\, v'_n, v'_n \,\rangle = -c_{p'+1}^2 - ... - c_{p'+m}^2 \leq 0.$$

Thus, $b_1^2 + ... + b_p^2 = 0$; and hence each $b_i = 0$ for $i = 1, ..., p$. Then since $(v'_1, ..., v'_n)$ is a basis and since the left-hand side of (5.13) is zero, $c_{p'+1} = ... = c_n = 0$. Therefore, the vectors in (5.12) are linearly independent, and this finishes the whole proof. □

NOTATION 5.3. *For dealing with indefinite forms, the notation $I_{p,m}$ is often used to denote the diagonal matrix*

$$I_{p,m} = \begin{bmatrix} I_p & \\ & -I_m \end{bmatrix}. \tag{5.14}$$

With this notation, the matrix representing the Lorentz form (5.10) is $I_{3,1}$.

Exercises for Section 5.2

EXERCISE 5.8. *Prove that a positive definite bilinear form is nondegenerate.*

EXERCISE 5.9. *A matrix A is called* ***positive-semidefinite*** *if $X^t A X \geq 0$ for each $X \in \mathbb{R}^n$. Prove that $A^t A$ is positive semidefinite for any $m \times n$ real matrix A.*

EXERCISE 5.10. *Find an orthogonal basis for the form on $\mathbb{R}^n$ whose matrix is as follows:*

(1) $\begin{bmatrix} 1 & 1 \\ 1 & 1 \end{bmatrix}$,

(2) $\begin{bmatrix} 1 & 0 & 1 \\ 0 & 2 & 1 \\ 1 & 1 & 1 \end{bmatrix}$.

(3) *Extend the vector* $X_1 = \frac{(1,1,1)^t}{\sqrt{3}}$ *to an orthonormal basis for* $\mathbb{R}^3$.

(4) *Prove that the only real matrix that is orthogonal, symmetric, and positive definite is the identity matrix.*

EXERCISE 5.11. *Let* W *be a subspace of a vector space* V *on which a symmetric bilinear form is given.*

(1) *Prove that* $W^\perp$ *is a subspace of* V.

(2) *Prove that the null space* N *is a subspace of* V.

EXERCISE 5.12. *Let* W_1, W_2 *be subspaces of a vector space* V *with a symmetric bilinear form. Prove each of the followings:*

(1) $(W_1 + W_2)^\perp = W_1^\perp \cap W_2^\perp$,

(2) $W_i \subset W_i^{\perp\perp}$,

(3) *If* $W_1 \subset W_2$ *then* $W_1^\perp \supset W_2^\perp$.

3. The Geometry Associated to a Symmetric Positive Form

In this section we go back to once more a symmetric positive definite bilinear form $\langle\ ,\ \rangle$ on an n-dimensional real vector space V.

DEFINITION 5.12. *A real n-dimensional vector space together with a symmetric positive definite bilinear form* $\langle\ ,\ \rangle$ *is called a* ***Euclidean space.***

One important consequence of the fact that the form is positive definite is that we can decide whether a vector v is zero or not by computing $\langle\ v, v\ \rangle$: $v = 0$ if and only if $\langle\ v, v\ \rangle = 0$, and this allows us to define

DEFINITION 5.13. *The* ***length*** *of a vector* v *is defined by the rule*

$$|v| = \sqrt{\langle\ v,\ v\rangle}, \tag{5.15}$$

in analogy with the length of vectors in $\mathbb{R}^n$ *[Cf. (4.15) in Chapter 4].*

We have seen that there is an orthonormal basis $\mathbf{B} = (v_1, ..., v_n)$ for V (Theorem 5.1), and with respect to this basis the form corresponds to dot product in $\mathbb{R}^n$:

$$\langle\, v, w \,\rangle = X^t Y,$$

where $v = \mathbf{B}X$ and $w = \mathbf{B}Y$. Using this correspondence we can transfer the geometry of $\mathbb{R}^n$ over to V. Thus, whenever a problem is presented to us on a Euclidean space V, a natural procedure will be to choose a convenient orthonormal basis by Gram-Schmidt procedure in the proof of Theorem 5.1, thereby reducing the problem to the familiar case of dot product on $\mathbb{R}^n$.

When a subspace W of V is given to us, there are two operations we can naturally make. The first is the restriction of the form to the subspace and the second is the projection of V onto W.

The restriction of the form to W is done by defining the value of the form on a pair w_1, w_2 of vectors in W to be $\langle\, w_1, w_2 \,\rangle$. Then it is a bilinear form on W and if the form on V is symmetric or if the form on V is symmetric and positive definite, then so is the restriction.

Restriction of the form can be used to define the unoriented angle between two vectors v, w. If the vectors are linearly dependent, then the angle is zero *by definition.* Otherwise, (v, w) is a basis of a two-dimensional subspace W of V. The restriction of the form to W is still positive definite, and hence there is an orthonormal basis (w_1, w_2) for W. By means of this basis, the vectors v, w correspond to their coordinate vectors X, Y in $\mathbb{R}^2$. This allows us to interpret geometric properties of the vectors v, w in terms of properties of X, Y.

Since the basis (w_1, w_2) is orthonormal, the form corresponds to dot product on $\mathbb{R}^2$: $\langle\, v, w \,\rangle = X^t Y$. Therefore, we have

$$|v| = |X|\,, |w| = |Y|\,, \text{ and } \langle\, v, w \,\rangle = X \cdot Y.$$

We *define* the angle θ between v and w to be the angle between X and Y, and thereby we obtain the formula

$$\langle\, v, w \,\rangle = |v|\,|w| \cos\theta, \tag{5.16}$$

as a consequence of the analogous formula (4.16) in Chapter 4 for dot product in $\mathbb{R}^2$.

This formula determines $\cos\theta$ in terms of the other symbols, and $\cos\theta$ determines θ up to a factor of ± 1. Therefore, the angel between v and w is determined up to sign by the form alone, and this is the best we can do, even when we are in $\mathbb{R}^3$.

Standard facts such as the *Schwarz Inequality*

$$|\langle\, v, w \,\rangle| \le |v|\,|w|\,, \tag{5.17}$$

and the *Triangle Inequality*

$$|v + w| \le |v| + |w| \tag{5.18}$$

can also be proved for arbitrary Euclidean spaces by restriction to a two-dimensional subspace. (Cf. For instance, Lemma 4.8 in Chapter 4).

The second operation is the projection of V onto W. Since the restriction of the form to W is positive definite, it is nondegenerate, and hence $V = W \oplus W^{\perp}$ by Proposition 5.6. Therefore, every $v \in V$ has a unique expression

$$v = w + w', \text{ with } w \in W \text{ and } \langle\, v, w' \,\rangle = 0. \tag{5.19}$$

DEFINITION 5.14. *The* ***orthogonal projection*** $\pi\colon V \longrightarrow W$ *is defined to be the linear transformation*

$$v \mapsto \pi(v) = w,$$

where w is as in (5.19).

An important fact is that the projected vector $\pi(v)$ can be computed easily in terms of an orthonormal basis $(w_1, w_2, ..., w_r)$ of W.

PROPOSITION 5.7. *Let* $(w_1, ..., w_r)$ *be an orthonormal basis of a subspace* W*, and let* $v \in V$*. Then the orthogonal projection* $\pi(v)$ *of* v *onto* W *is the vector*

$$\pi(v) = \langle\, v, w_1 \,\rangle w_1 + ... + \langle\, v, w_r \,\rangle w_r.$$

PROOF. We denote the right side of the equality by $\tilde{w}$. Then since $\langle\, \tilde{w}, w_i \,\rangle = \langle\, v, w_i \,\rangle\langle\, w_i, w_i \,\rangle = \langle\, v, w_i \,\rangle$ for each $i = 1, ..., r$, we see that $v - \tilde{w} \in W^{\perp}$. Thus, we have $v = \tilde{w} + (v - \tilde{w})$ and this expression is unique by (5.19) above, i.e., $w = \tilde{w}$ and $w' = v - \tilde{w}$ there. □

REMARK 5.10. *Thus if* $\pi(v)$ *is defined as* $\pi(v) = \langle\, v, w_1 \,\rangle w_1 + ... + \langle\, v, w_r \,\rangle w_r$*, then* $v - \pi(v)$ *is orthogonal to* W*. We note that this formula explains the geometric meaning of the Gram-Schmidt orthogonalization procedure described in the proof of Theorem 5.1.*

When $W = V$ with the identity map for π, we have an important description for coefficients:

COROLLARY 5.7. *Let* $\mathbf{B} = (w_1, ..., w_n)$ *be an orthonormal basis for a Euclidean space* V*. Then*

$$v = \langle\, v, w_1 \,\rangle w_1 + ... + \langle\, v, w_r \,\rangle w_r + ... + \langle\, v, w_n \,\rangle w_n.$$

In other words, the coordinate vector in $\mathbb{R}^n$ *of* v *with respect to the orthonormal basis* $\mathbf{B}$ *is*

$$X = (\langle\, v, w_1 \,\rangle, ..., \langle\, v, w_n \,\rangle)^t. \qquad \square$$

Exercises for Section 5.3

EXERCISE 5.13. *Let* V *be a Euclidean space. Prove the Schwarz Inequality and the Triangle Inequality.*

EXERCISE 5.14. *Let* W *be a subspace of a Euclidean space* V*. Prove that* $W = W^{\perp\perp}$*.*

EXERCISE 5.15. *Let* V *be a Euclidean space. Show that if* $|v| = |w|$ *then* $(v + w) \perp (v - w)$*. Can you interpret this formula geometrically?*

EXERCISE 5.16. *Prove the* ***parallelogram law,***

$$|v+w|^2 = |v-w|^2 = 2\,|v|^2 + 2\,|w|^2,$$

in a Euclidean space.

EXERCISE 5.17. *Prove that the orthogonal projection is a linear transformation.*

EXERCISE 5.18. *Let A be a real symmetric matrix, and let T be a linear operator on $\mathbb{R}^n$ whose matrix is A.*

(1) *Prove that* $\ker(T) \perp im(T)$ *and that* $V = \ker(T) \oplus im(T)$.
(2) *Prove that T is an orthogonal projection onto $im(T)$ if and only if, in addition to being symmetric, $A^2 = A$.*

4. Hermitian Forms

In this section, we assume that our scalar field is the field $\mathbb{C}$ of complex numbers.

In working with complex vector spaces, since $\mathbb{C}$ is identified with $\mathbb{R}^2$ we can consider a complex vector $v \in \mathbb{C}^n$ as a real vector $v \in \mathbb{R}^{2n}$ and we can consider its length as follows:

DEFINITION 5.15. *If $X = (x_1, ..., x_n)^t$ is a complex vector with each $x_r = a_r + b_r i$, where $i = \sqrt{-1}$, the* ***length*** *of X is*

$$\begin{aligned} |X| &= \sqrt{(a_1^2 + b_1^2) + ... + (a_n^2 + b_n^2)} \\ &= \sqrt{x_1\bar{x}_1 + ... + x_n\bar{x}_n}, \end{aligned} \tag{5.20}$$

where the bar denotes complex conjugation: $x_r = a_r + b_r i \mapsto \bar{x}_r = a_r - b_r i$.

Thus, under any identification of $\mathbb{C}^n$ with $\mathbb{R}^{2n}$ we can still consider the *same-sized* length of a complex vector $c \in \mathbb{C}^n$ defined as above. But, if we attempt to use the usual dot product formula for vectors in $\mathbb{R}^{2n}$ to express the formula we obviously fails in reflecting this important point. Nevertheless, since the mapping called "complex conjugation" fixes any real numbers, we can safely modify the dot product as follows:

Definition 5.16. *The **standard hermitian product**, or the **hermitian dot product** of two complex vectors $X = (x_1, ..., x_n)^t$ and $Y = (y_1, ..., y_n)^t$ is defined by the formula*

$$\langle X, Y \rangle = \bar{X}^t Y = \bar{x}_1 y_1 + ... + \bar{x}_n y_n. \tag{5.21}$$

We first note that the formula in (5.21) agrees with dot product for real vectors. We also note that it obviously satisfies the following properties.

Lemma 5.3.

a: *(Positivity Property)*

$$\langle X, X \rangle \text{ is a positive real number if } X \neq 0. \tag{5.22}$$

b: *(Linearity in the second variable)* $\langle X, cY \rangle = c\langle X, Y \rangle$ *and* $\langle X, Y_1 + Y_2 \rangle = \langle X, Y_1 \rangle + \langle X, Y_2 \rangle$.

c: *(Conjugate linearity in the first variable)* $\langle cX, Y \rangle = \bar{c}\langle X, Y \rangle$ *and* $\langle X_1 + X_2, Y \rangle = \langle X_1, Y \rangle + \langle X_2, Y \rangle$

d: *(Hermitian symmetry)* $\langle Y, X \rangle = \overline{\langle X, Y \rangle}$. □

Thus, we have a positive definite product on $\mathbb{C}^n$ at a small cost in linearity and symmetry. More generally, we have the following definition:

Definition 5.17. *A **hermitian form** on a complex vector space V is a function of two variables to the field $\mathbb{C}$ of complex numbers*

$$\begin{aligned} V \times V &\longrightarrow \mathbb{C} \\ (v, w) &\longmapsto \langle v, w \rangle \end{aligned} \tag{5.23}$$

*satisfying the relations in Lemma 5.3, except for part a. If in addition it satisfies, the positivity property, i.e., part a of Lemma 5.3, we call it a **positive definite hermitian form** on V.*

Example 5.7. *(5.21) is thus an example of a positive definite hermitian form on a complex vector space $\mathbb{C}^n$.*

In what follows we are going to define several terminologies which are in most cases generalization of those defined in the case of real vector spaces. We are also going to extend several results in the case of real vector spaces in order to make them sense for complex vector space. Since our approach is almost the same as the previous case of real vector spaces, this is done mostly without providing proof.

DEFINITION 5.18. *Let* $\mathbf{B} = (v_1, ..., v_n)$ *be a basis for* V. *Then the* ***matrix*** *of the form is defined in the analogous way as the matrix of a bilinear form in (5.7):*

$$A = (a_{ij}), \qquad \textit{where } a_{ij} := \langle\, v_i, v_j \,\rangle.$$

Then the formula for the form now becomes

$$\langle\, v, w \,\rangle = \overline{X}^t AY, \tag{5.24}$$

if $v = \mathbf{B}X$ and $w = \mathbf{B}Y$. (Cf. (5.7)).

Because the form is hermitian symmetric, we immediately see that

$$a_{ij} = \langle\, v_i, v_j \,\rangle = \overline{\langle\, v_j, v_i \,\rangle} = \overline{a}_{ji},$$

that is, $A = \overline{A}^t$.

DEFINITION 5.19. *The matrix* $\overline{A}^t$ *is called the* ***adjoint*** *of a matrix* A *and is denoted by* A^* *[different from the one defined in Chapter 2]:*

$$A^* = \overline{A}^t.$$

LEMMA 5.4.

a: $(A + B)^* = A^* + B^*$.
b: $(AB)^* = B^*A^*$.
c: $(A^*)^{-1} = (A^{-1})^*$.
d: $A^{**} = A$.

The proof of this lemma is very easy. We note that if A is a real matrix, then A^* is the same as A^t.

Then formula (5.24) can be rewritten as

$$\langle\, v, w \,\rangle = X^*AY, \tag{5.25}$$

and the standard hermitian product on $\mathbb{C}^n$ becomes $\langle\, X, Y \,\rangle = X^*Y$.

DEFINITION 5.20. *A matrix is called* ***hermitian*** *or* ***self-adjoint*** *if*

$$A = A^*. \tag{5.26}$$

Thus, the matrix A associated to a hermitian form on a complex vector space V is hermitian. By definition, the diagonal entries of a hermitian matrix A must be real numbers and the entries of A below the diagonal are complex conjugates of those above the diagonal:

$$A = \begin{bmatrix} r_1 & & & & \\ & \cdot & & a_{ij} & \\ & & \cdot & & \\ & \bar{a}_{ij} & & \cdot & \\ & & & & r_n \end{bmatrix}, \qquad r_i \in \mathbb{R}, \quad a_{ij} \in \mathbb{C}.$$

Thus,

(5.27) *The real hermitian matrices are the real symmetric matrices.*

In Sections 1 and 2 of this chapter, we have discussed change of basis. Here we also have its analogues for hermitian forms. Given a hermitian form, if P is the transition matrix from an old basis to a new basis, as is explained in (5.9) a change of basis by P leads to

$$X'^* AY' = (PX)^* A' PY = X^*(P^* A' P)Y,$$

and hence the new matrix A' satisfies

$$A = P^* A' P \quad \text{or} \quad A' = (P^*)^{-1} A P^{-1}. \tag{5.28}$$

Then replacing P with $Q = (P^*)^{-1}$, we obtain the following result that is analogous to Corollary 5.2:

COROLLARY 5.8. *Let A be the matrix of a hermitian form with respect to a basis. Those matrices representing the same hermitian form with respect to different bases are of the form*

$$A' = QAQ^*,$$

for some invertible matrix $Q \in GL_n(\mathbb{C})$. □

For hermitian forms, the analogues of orthogonal matrices are the unitary matrices:

DEFINITION 5.21. *A matrix P is called **unitary** if*

$$P^{-1} = P^*. \tag{5.29}$$

EXAMPLE 5.8. *We easily check that*

$$\frac{1}{\sqrt{2}}\begin{bmatrix} 1 & i \\ 1 & -i \end{bmatrix}$$

is a unitary matrix.

Thus, we see that

(5.30) *The real unitary matrices are the real orthogonal matrices.*

We note that if P is a unitary matrix, then so is its inverse.

Moreover, formula (5.28) says that unitary matrices represent those changes of basis that leave the standard hermitian product X^*Y invariant under change of basis.

COROLLARY 5.9. *A change of basis preserves the standard hermitian product, that is, $X^*Y = X'^*Y'$, if and only if its transition matrix P is unitary.* □

The proof of this corollary is very similar to that of Corollary 5.1.

In general, Corollary 5.8 tells us that a general change of basis changes the standard hermitian product X^*Y to $X'^*A'Y'$, where $A' = QQ^*$ for some $Q \in GL_n(\mathbb{C})$.

The notion of orthogonality for hermitian forms is defined exactly as for symmetric bilinear forms:

DEFINITION 5.22. *Given a hermitian form $\langle\, v, w \,\rangle$ on a complex vector space V, a vector $v \in V$ is **orthogonal** to $w \in V$ if $\langle\, v, w \,\rangle = 0$.*

Then since the form is hermitian symmetric, orthogonality is still a symmetric relation.

Because of these, we can now copy the discussion of Sections 1 and 2 for hermitian forms without essential change, and Sylvester's Law (Theorem 5.5) for real symmetric forms carries over to the hermitian case.

In particular, we can also speak of positive definite forms and of orthonormal bases:

DEFINITION 5.23. *A hermitian form* $\langle\ ,\ \rangle$ *on a complex vector space* V *is* ***positive definite*** *if*

$$\langle\, v, v\,\rangle \text{ is a positive real number if } v \neq 0. \tag{5.31}$$

A basis $B = (v_1, ..., v_n)$ *for a complex vector space* V *with a hermitian form* $\langle\ ,\ \rangle$ *on it is* ***orthonormal*** *if*

$$\langle\, v_i, v_i\,\rangle = 1 \quad \text{and} \quad \langle\, v_i, v_j\,\rangle = 0 \ \text{ if } i \neq j. \tag{5.32}$$

THEOREM 5.6. *Let* $\langle\ ,\ \rangle$ *be a hermitian form on a complex vector space* V*. There exists an orthonormal basis for* V *if and only if the form is positive definite.* □

This theorem is true by (5.32) and by a similar argument of the proof of Theorem 5.1.

PROPOSITION 5.8. *Let* W *be a subspace of a complex vector space* V *with a hermitian form on it. If the restriction of the form to* W *is nondegenerate, then* $V = W \oplus W^{\perp}$. □

The proof of this proposition is almost the same as the one for Proposition 5.6.

Exercises for Section 5.4

EXERCISE 5.19. *Prove Lemma 5.3.*

EXERCISE 5.20. *Show that the dot product form* $X \cdot Y = X^t Y$ *is not positive definite on* $\mathbb{C}^n$.

EXERCISE 5.21. *Prove that a matrix* A *is hermitian if and only if the associated form* X^*AX *is a hermitian form.*

EXERCISE 5.22. *Prove that if X^*AX is real for all complex vectors X, then A is hermitian.*

EXERCISE 5.23. *Let $X, Y \in \mathbb{C}^n$ be orthonormal bases. Prove that*

$$|X+Y|^2 = |X|^2 + |Y|^2.$$

EXERCISE 5.24. *Is $\langle X, Y \rangle = x_1y_1 + ix_1y_2 - ix_2y_1 + ix_2y_2$, where $X = (x_1, x_2)^t$ and $Y = (y_1, y_2)^t$, a hermitian form?*

EXERCISE 5.25. *Prove that the determinant of a hermitian matrix is a real number.*

EXERCISE 5.26. *Prove that A is a positive definite hermitian if and only if $A = P^*P$ for an invertible matrix P.*

EXERCISE 5.27. *Let A be a unitary matrix. Prove that*

$$|\det(A)| = 1.$$

5. The Spectral Theorem

In this section we will study an n-dimensional complex vector space V with a positive definite hermitian form $\langle\ ,\ \rangle$ on it.

DEFINITION 5.24. *A complex vector space V on which a positive definite hermitian form is given is called a* ***hermitian space****.*

We note that, by Theorem 5.6, we can change any basis for V to an orthonormal basis, and consequently, with respect to this orthonormal basis, we can identify V with $\mathbb{C}^n$ with the standard hermitian product X^*Y. Therefore, we will assume in what follows that our bases are orthonormal, once the form is given to us.

This changes all previous calculations in the following way: If $\mathbf{B} = (v_1, ..., v_n), \mathbf{B}' = (v_1', ..., v_n')$ are two *orthonormal* bases, then the transition matrix P relating them will be unitary: The fact that the bases are orthonormal means that the matrix of the form $\langle\ ,\ \rangle$ with respect to each basis is the identity matrix I, and hence (5.28) says that $I = P^*IP$, or $P^*P = I$. Thus

LEMMA 5.5. *For a fixed orthonormal basis* $\mathbf{B}$, *a new basis* $\mathbf{B}'$ *is orthonormal if and only if the invertible matrix* P *is unitary, where* $\mathbf{B} = \mathbf{B}'P$. □

Having this in our mind, we are now going to study a linear operator

$$T\colon V \longrightarrow V \tag{5.33}$$

on a hermitian space V. Let $\mathbf{B}$ be an orthonormal basis of V, and let M be the associated matrix of T with respect to $\mathbf{B}$. If we change from the old basis $\mathbf{B}$ to a new basis $\mathbf{B}'$, then the matrix M is changed to $M' = PMP^{-1}$, where P is the transition matrix from $\mathbf{B}$ to $\mathbf{B}'$, computed from $\mathbf{B} = \mathbf{B}'P$. On the other hand, since both bases $\mathbf{B}$ and $\mathbf{B}'$ are orthonormal, the matrix P must be unitary as we already have seen it above; hence

$$M' = PMP^* \tag{5.34}$$

PROPOSITION 5.9. *Let* T *be a linear operator on a hermitian space* V, *and let* $M = (b_{ij})$ *be the matrix of* T *with respect to an orthonormal basis* $\mathbf{B}$.

- **a:** The matrix M is hermitian if and only if $\langle\, v, Tw \,\rangle = \langle\, Tv, w \,\rangle$ for all $v, w \in V$, and if the operator T satisfies the property in the right side of this equivalence, T is called a **hermitian operator**.
- **b:** The matrix M is unitary if and only if $\langle\, v, w \,\rangle = \langle\, Tv, Tw \,\rangle$ for all $v, w \in V$, and if the operator T satisfies the property in the right side of this equivalence, T is called a **unitary operator**.

PROOF. Let X, Y be the coordinate vectors of v, w: $v = \mathbf{B}X, w = \mathbf{B}Y$, so that $\langle\, v, w \,\rangle = X^*Y$ and $Tv = \mathbf{B}MX$.

Part a: Under these settings, we have

$$\langle\, v, Tw \,\rangle = X^*MY, \text{ and } \langle\, Tv, w \,\rangle = (MX)^*Y = X^*M^*Y.$$

Thus if $M = M^*$, then $\langle\, v, Tw \,\rangle = \langle\, Tv, w \,\rangle$ for all $v, w \in V$, i.e., T is hermitian. Conversely, if T is hermitian, we particularly choose $v = e_i$, $w = e_j$ to interpret both sides of the equality: $\langle\, v, Tw \,\rangle = \langle\, Tv, w \,\rangle$: $LHS = e_i^*Me_j = b_{ij}$ and $RHS = e_i^*M^*e_j = \overline{b}_{ji}$. Thus, M is hermitian.

We recall that we have used the same method in the proof of Proposition 5.1.

Part b: Under the same settings, we have

$$\langle\, v, w \,\rangle = X^*Y, \text{ and } \langle\, Tv, Tw \,\rangle = X^*M^*MY.$$

Thus $\langle\, v, w \,\rangle = \langle\, Tv, Tw \,\rangle$ for all $v, w \in V$ if and only if $MM^* = I$. □

THEOREM 5.7. *(Spectral Theorem)*

a: Let T be a hermitian operator on a hermitian vector space V. Then there exists an orthonormal basis of V consisting of eigenvectors of T.

b: (Matrix form) Let M be a hermitian matrix with respect to an orthonormal basis. Then there exists a unitary matrix P such that PMP^* is a *real* diagonal matrix.

PROOF. We first prove that part a and part b are equivalent. If part a holds, then we have a change of basis from a given old orthonormal basis $\mathbf{B}$ to a new orthonormal basis $\mathbf{B}'$ consisting of eigenvectors of T. Let M be the hermitian matrix of T with respect to $\mathbf{B}$. We already know by Lemma 5.5 that the transition matrix P from $\mathbf{B}$ to $\mathbf{B}'$ is unitary, and hence the matrix $M' = PMP^*$ of T with respect to the new basis is again hermitian. Thus the diagonal entries of M' are all real. On the other hand, since $\mathbf{B}'$ consists of eigenvectors v_i of T, $T(v_i) = \lambda_i v_i$ for each i, and since this determines the whole entries of the matrix M', part b holds.

Conversely, if part b holds, then $\mathbf{B}' = \mathbf{B}P^* = (v_1, ..., v_n)$ is a new orthonormal basis, where $\mathbf{B}$ is the old orthonormal basis with respect to which the matrix of T is M. This is so by Lemma 5.5. Then since the matrix of T with respect to $\mathbf{B}'$ is the diagonal matrix $M' = PMP^*$ with (real) diagonal entries λ_i, each $v_i \in \mathbf{B}'$ must be an eigenvector of T, and moreover they must be already normalized since we have already seen that $\mathbf{B}'$ is an orthonormal basis. Thus, part a holds.

Now, we are going to prove part b. We recall that the characteristic polynomial of M over the field $\mathbb{C}$ of complex numbers has a root $a_1 \in \mathbb{C}$;

hence its eigenspace $E_{a_1} \supset \{0\}$; hence there exists an eigenvector $v = v_1$ of which eigenvalue is a_1. We normalize it so that its length is 1: $\langle\, v, v \,\rangle = 1$. We then extend it to a basis of V; and then we get an orthonormal basis $\mathbf{B}'$ of V by Theorem 5.6. Then the matrix M' of T with respect to $\mathbf{B}'$ looks like

$$M_1 = P_1 M P_1^* = \begin{bmatrix} a_1 & * & \cdot & \cdot & \cdot & * \\ 0 & \lceil & & & & \rceil \\ \cdot & & & & & \\ \cdot & & & N_2 & & \\ \cdot & & & & & \\ 0 & \lfloor & & & & \rfloor \end{bmatrix}$$

for some unitary matrix P_1 by Lemma 5.5, where N_2 is an $(n-1) \times (n-1)$ submatrix. Since T is hermitian, so is the matrix M_1 by Proposition 5.9. Thus, a_1 is real, the rest entries of the first row are all zeros, and N_2 is again hermitian. Thus, we proceed by induction on n, and it goes as follows: By induction hypothesis, there exists a unitary matrix P_2 of size $(n-1) \times (n-1)$ such that

$$P_2' N_2 P_2'^* = \begin{bmatrix} a_2 & & & \\ & \cdot & & \\ & & \cdot & \\ & & & \cdot \\ & & & & a_n \end{bmatrix},$$

with real diagonal entries a_i. Then

$$P_2 = \begin{bmatrix} 1 & 0 \\ 0 & P_2' \end{bmatrix}$$

is clearly unitary; hence so is $P_2 P_1$. Therefore,

$$M_2 = P_2 P_1 M P_1^* P_2^* = \begin{bmatrix} a_1 & & & & \\ & a_2 & & & \\ & & \cdot & & \\ & & & \cdot & \\ & & & & a_n \end{bmatrix}$$

is the desired real diagonal matrix. □

REMARK 5.11. *To diagonalize a hermitian matrix M (with respect to the standard basis* $\mathbf{E}$*) by a unitary matrix P, one can proceed first by determining the eigenvalues. If the eigenvalues are all distinct, then the corresponding eigenvectors are going to be orthogonal. This follows from the Spectral Theorem (Theorem 5.7) above. Let* $\mathbf{B}'$ *be the orthonormal basis obtained by normalizing the lengths of the eigenvectors to* 1*. Then, the transition matrix* $P = \mathbf{B}'^{-1}$ *is going to be automatically unitary.*

EXAMPLE 5.9. *Let* $M = \begin{bmatrix} 2 & i \\ i & 2 \end{bmatrix}$. *One easily computes that the eigenvalues of this matrix are real numbers* $3, 1$, *and one can also see that the vectors*

$$v_1' = \begin{bmatrix} 1 \\ -i \end{bmatrix}, v_2' = \begin{bmatrix} 1 \\ i \end{bmatrix}$$

are eigenvectors with these eigenvalues. We then normalize their length to 1 *by the factor* $\frac{1}{\sqrt{2}}$ *to get another basis* $\mathbf{B}' = (\frac{1}{\sqrt{2}}\begin{bmatrix} 1 \\ -i \end{bmatrix}, \frac{1}{\sqrt{2}}\begin{bmatrix} 1 \\ i \end{bmatrix})$ *again consisting of eigenvectors with these eigenvalues and, since we have an equation* $\mathbf{E} = \mathbf{B}'P$ *for* P, *we get:*

$$P^{-1} = P^* = \mathbf{B}'.$$

Thus

$$\begin{aligned} P &= (P^*)^* = \mathbf{B}'^* \\ &= \frac{1}{\sqrt{2}}\begin{bmatrix} 1 & 1 \\ -i & i \end{bmatrix}^* = \frac{1}{\sqrt{2}}\begin{bmatrix} 1 & i \\ 1 & -i \end{bmatrix} \end{aligned}$$

and hence

$$PMP^* = \begin{bmatrix} 3 & 0 \\ 0 & 1 \end{bmatrix}.$$

REMARK 5.12. *But the Spectral Theorem (Theorem 5.7) also asserts that a hermitian matrix can be diagonalized* even if eigenvalues aren't distinct.

Here is an interesting corollary of the Spectral Theorem, which we can give a direct proof.

PROPOSITION 5.10. *The eigenvalues of a hermitian operator T on a hermitian space V are real numbers.*

PROOF. Let a be an eigenvalue and $v \neq 0$ be its associated eigenvector for T, i.e., $T(v) = av$. Then by Proposition (5.9) $\langle\ v, Tv\ \rangle = \langle\ Tv, v\ \rangle$ holds, and hence $\langle\ v, av\ \rangle = \langle\ av, v\ \rangle$. Then by conjugate linearity in Lemma 5.3, we see that

$$a\langle\ v, v\ \rangle = \bar{a}\langle\ v, v\ \rangle.$$

But, $\langle\ v, v\ \rangle \neq 0$ since the form is positive definite. Therefore, $a = \bar{a}$, i.e., a is a real number. □

The results we have proved for hermitian matrices have analogues for real symmetric matrices.

Let V be a real vector space with a positive definite bilinear form $\langle\ ,\ \rangle$ on it and let T be a linear operator on V. Under these settings we have:

PROPOSITION 5.11. *Let M be the matrix of T with respect to an orthonormal basis.*

- **a:** *The matrix M is symmetric if and only if $\langle\ v, Tw\ \rangle = \langle\ Tv, w\ \rangle$ for all $v, w \in V$, and if so, T is called a* ***symmetric operator****.*
- **b:** *The matrix M is orthogonal if and only if $\langle\ v, w\ \rangle = \langle\ Tv, Tw\ \rangle$ for all $v, w \in V$, and if so, T is called an* ***orthogonal operator****.* □

PROPOSITION 5.12. *The eigenvalues of a real symmetric matrix are real.*

PROOF. A real symmetric matrix is hermitian; hence this is a special case of Proposition 5.10. □

THEOREM 5.8. *(Spectral Theorem (real case))*

a: Let T be a symmetric operator on a real vector space with a positive definite bilinear form on it. Then there exists an orthonormal basis consisting of eigenvectors of T.

b: (Matrix form) Let M be a real symmetric $n \times n$ matrix with respect to an orthonormal basis. Then there exists an orthogonal matrix $P \in GL_n(\mathbb{R})$ such that PMP^t is diagonal.

PROOF. Since we now know that the eigenvalues of such an operator are real, we can copy the proof of Theorem 5.7. □

Exercises for Section 5.5

EXERCISE 5.28. *Prove that the eigenvalues of a real symmetric matrix are real numbers.*

EXERCISE 5.29. *Prove that eigenvectors associated to distinct eigenvalues of a hermitian matrix A are orthogonal.*

EXERCISE 5.30. *Find a unitary matrix P so that PAP^* is diagonal, where*

$$A = \begin{bmatrix} 1 & i \\ -i & 1 \end{bmatrix}.$$

EXERCISE 5.31. *Find a real orthogonal matrix P such that PAP^t is diagonal, where*

(1) $A = \begin{bmatrix} 1 & 2 \\ 2 & 1 \end{bmatrix}$,

(2) $A = \begin{bmatrix} 1 & 1 & 1 \\ 1 & 1 & 1 \\ 1 & 1 & 1 \end{bmatrix}$,

(3) $A = \begin{bmatrix} 1 & 0 & 1 \\ 0 & 1 & 0 \\ 1 & 0 & 0 \end{bmatrix}$.

EXERCISE 5.32. *Show that the only matrix that is both positive definite hermitian and unitary is the identity matrix I.*

EXERCISE 5.33. *Prove that for any square matrix A,* $\ker(A) = (im(A^*))^{\perp}$.

EXERCISE 5.34. *Prove that for every complex matrix A there exists a unitary matrix P such that PAP^* is upper triangular.*

EXERCISE 5.35. *Let A be a hermitian matrix. Prove that there exists a unitary matrix P with determinant 1 such that PAP^* is diagonal.*

6. The Spectral Theorem for Normal Operators

The Spectral Theorem 5.7 tells us that any hermitian matrix M can be transformed into a real diagonal matrix C by a unitary matrix P: $D = PMP^*$. We now ask for the matrices M which can be transformed in the same way to a diagonal matrix D, but this time we no longer require D to be a real matrix. It turned out that there is an elegant formal characterization of such matrices.

DEFINITION 5.25. *A matrix M is called* ***normal*** *if it commutes with its adjoint, that is, $MM^* = M^*M$.*

EXAMPLE 5.10. *Every hermitian matrix is normal; moreover every unitary matrix is normal.*

LEMMA 5.6. *Let P be a unitary matrix. If M is normal, then $M' = PMP^*$ is also normal, and conversely.*

PROOF. If M is normal, then

$$\begin{aligned} M'M'^* &= PMP^*(PMP^*)^* = PMP^*(PM^*P^*) \\ &= PMM^*P^* = PM^*MP^* \\ &= PM^*(P^*P)MP^* = (PMP^*)^*(PMP^*) \\ &= M'^*M'. \end{aligned}$$

Its converse is very easy. □

Since this lemma says that normality is preserved under any change of basis by a unitary matrix P from an old orthonormal basis to a new orthonormal basis, this lemma allows us to define:

DEFINITION 5.26. *A* ***normal operator*** $T\colon V \longrightarrow V$ *on a hermitian space* V *is a linear operator whose matrix* M *with respect to any orthonormal basis is a normal matrix.*

THEOREM 5.9. *A complex matrix* M *is normal if and only if there exists a unitary matrix* P *such that* PMP^* *is diagonal.*

PROOF. We first note that a diagonal matrix is normal because any two diagonal matrices commute: $DD^* = D^*D$. Then by Lemma 5.6 M is normal, and this proves the $\Leftarrow$ direction.

Conversely, suppose that M is normal. Choose an eigenvector $v = v_1$ of M, and normalize so that $\langle\ v, v\ \rangle = 1$, as is done at the beginning of the proof of Theorem 5.7. Extend $\{v_1\}$ to a basis of V; then further to an orthonormal basis of V. Then M will be changed to a normal matrix

$$M_1 = P_1 M P_1^* = \begin{bmatrix} a_{11} & a_{12} & \cdot & \cdot & \cdot & a_{1n} \\ 0 & \lceil & & & & \rceil \\ \cdot & & & & & \\ \cdot & & & N_2 & & \\ \cdot & & & & & \\ 0 & \lfloor & & & & \rfloor \end{bmatrix}$$

for some unitary matrix P_1, and hence

$$M_1^* = P_1 M^* P_1^* = \begin{bmatrix} a_{11} & 0 & \cdot & \cdot & \cdot & 0 \\ \bar{a}_{12} & \lceil & & & & \rceil \\ \cdot & & & & & \\ \cdot & & & N_2^* & & \\ \cdot & & & & & \\ \bar{a}_{1n} & \lfloor & & & & \rfloor \end{bmatrix}.$$

Now the $(1,1)$ entry of $M_1^* M_1$ is $a_{11}\bar{a}_{11}$, while the same entry of $M_1 M_1^*$ is $a_{11}\bar{a}_{11} + a_{12}\bar{a}_{12} + ... + a_{1n}\bar{a}_{1n}$. Since M is normal, so is M_1, that is $M_1^* M_1 = M_1 M_1^*$, it follows that $a_{12}\bar{a}_{12} + ... + a_{1n}\bar{a}_{1n} = 0$. Then since

$a_{1j}\bar{a}_{1j} \geq 0$ for $j \geq 1$, this shows that $a_{1j} = 0$ for $j > 1$; hence

$$M_1 = P_1 M P_1^* = \begin{bmatrix} a_{11} & 0 & \cdot & \cdot & \cdot & 0 \\ 0 & \lceil & & & & \rceil \\ \cdot & & & & & \\ \cdot & & & N_2 & & \\ \cdot & & & & & \\ 0 & \lfloor & & & & \rfloor \end{bmatrix}.$$

We then continue, working on N_2, i.e., by induction on n, similar to what we have done in the proof of Theorem 5.7 because N_2 is normal. □

Exercises for Section 5.6

EXERCISE 5.36. *Prove that for any normal matrix A, $\ker(A) = (im(A))^\perp$.*

EXERCISE 5.37. *Let P be a real matrix which is normal and has real eigenvalues. Prove that P is symmetric.*

EXERCISE 5.38. *Let P be a real skew-symmetric matrix. Prove that P is normal.*

EXERCISE 5.39. *Let A be a normal matrix. Prove that A is hermitian if and only if all eigenvalues of A are real.*

EXERCISE 5.40. *Let A be a normal matrix. Prove that A is unitary if and only if every eigenvalues of A has absolute value 1.*

7. Conics and Quadrics

As an application of the theory we have so far developed in this Chapter, we are going to work on a concrete and classical example in this section, which is about the classification of nondegenerate conics in $\mathbb{R}^2$and that of nondegenerate quadrics in $\mathbb{R}^3$.

DEFINITION 5.27. *A **conic** is the locus in the plane $\mathbb{R}^2$ defined by a quadratic equation in two variables, of the form*

$$f(x_1, x_2) = a_{11}x_1^2 + 2a_{12}x_1x_2 + a_{22}x_2^2 + b_1x_1 + b_2x_2 + c = 0, \tag{5.35}$$

where a_{ii}, b_i, and c are real numbers. A conic is called **degenerate** *if it is a pair of lines, a single line, a point, or empty, depending on the particular equation. Otherwise, it is called* **nondegenerate.**

The term **quadric** is used to designate the analogous loci in three or more dimensions.

DEFINITION 5.28. *The quadratic part of $f(x_1, x_2)$ is called a* **quadratic form:**

$$q(x_1, x_2) = a_{11}x_1^2 + 2a_{12}x_1x_2 + a_{22}x_2^2. \tag{5.36}$$

In general, a **quadratic form** *in n variables, $x_1x_2, ..., x_n$ is a polynomial each of whose terms has degree 2 in the variables.*

Because of this, a quadratic form is called **homogeneous.**

It is convenient to express the form $q(x_1, x_2)$ in matrix notation. To do this, we introduce the real symmetric matrix.

$$A = \begin{bmatrix} a_{11} & a_{12} \\ a_{12} & a_{22} \end{bmatrix}. \tag{5.37}$$

Then $q(x_1, x_2) = X^tAX$, where $X = (x_1, x_2)^t$. We also introduce the row vector $B = (b_1, b_2)$. Then equation (5.35) can be written in matrix notation as

$$X^tAX + BX + c = 0. \tag{5.38}$$

In the above, we put 2 into formulas (5.35) and (5.36) in order to avoid some coefficient $\frac{1}{2}$ in the matrix (5.37). Thus, an alternative way to writ the quadratic form would be

$$q(x_1, x_2) = a_{11}x_1^2 + a_{12}x_1x_2 + a_{12}x_2x_1 + a_{22}x_2^2.$$

We now propose to describe the congruence classes of conics as geometric figures, or what is the same, their orbits under the action of the group M of rigid motions of the plane. (Cf. Definition 4.20). A rigid motion will produce a change of variable in equation (5.35).

THEOREM 5.10. *Every nondegenerate conic is congruent to one of the following: For* $a_{11}, a_{22} > 0$,

a: *(Ellipse)* $a_{11}x_1^2 + a_{22}x_2^2 - 1 = 0$,
b: *(Hyperbola)* $a_{11}x_1^2 - a_{22}x_2^2 - 1 = 0$,
c: *(Parabola)* $a_{11}x_1^2 - x_2 = 0$.

PROOF. We are going to simplify Equation (5.35) in two steps, first applying an orthogonal transformation (a rotation or reflection) to diagonalize its associated form A and then applying a translation to eliminate, as much as possible, the linear and constant terms $BX + c$.

By the Spectral Theorem (5.7), there exists an orthogonal matrix P such that PAP^t is diagonal. We make the change of variable, $X' = PX$, or $X = P^tX'$. Substituting it into Equation (5.38) gives

$$X'^t(PAP^t)X' + (BP^t)X' + c = 0. \tag{5.39}$$

Thus, there exists an orthogonal change of variable such that the quadratic form becomes diagonal, that is, the coefficient a_{12} of x_1x_2 is zero.

Now, suppose that A is diagonal. Then f has the form

$$f(x_1, x_2) = a_{11}x_1^2 + a_{22}x_2^2 + b_1x_1 + b_2x_2 + c = 0.$$

If both a_{11} and a_{22} are not zero, then we eliminate b_i by **completing the squares**, and then making the substitution

$$x_i = (x_i' - \frac{b_i}{2a_{ii}}). \tag{5.40}$$

This substitution results in

$$a_{11}x_1'^2 + a_{22}x_2'^2 + c' = 0, \tag{5.41}$$

where c' is a number which can be determined if desired. We note that $c' = 0$ gives a degenerate conic, a pair of lines, and hence we may assume that c' is not zero. Then by dividing both sides of the equality in (5.41) by c' and reflecting possible signs, we obtain either part a or part b.

Now suppose that $a_{11} \neq 0$ but $a_{22} = 0$. If $b_1 \neq 0$ then we first apply Equation (5.40) for $i = 1$ in order to get rid of the term b_1x_1. After that, if $b_2 = 0$ then we have a degenerate conic, a point; hence we may assume that $b_2 \neq 0$. We then use the substitution

$$x_2 = x_2' - c/b_2 \tag{5.42}$$

to eliminate the constant term c. We then normalize one coefficient to -1 to obtain part c. □

The method used above can be applied in any number of variables to classify quadrics in n dimensions. The general quadric equation has the form

$$f(x_1, ..., x_n) = \sum_i a_{ii} x_i^2 + \sum_{i<j} 2a_{ij} x_i x_j + \sum_i b_i x_i + c = 0, \tag{5.43}$$

where all coefficients are real numbers. We could also write this equation more compactly as

$$f(x_1, ..., x_n) = \sum_{i,j} a_{ij} x_i x_j + \sum_i b_i x_i + c = 0, \tag{5.44}$$

where the first sum is over all pairs of indices, and where we set $a_{ji} = a_{ij}$.

We define the matrices A, B to be

$$A = \begin{bmatrix} a_{11} & a_{12} & \cdot & \cdot & \cdot & a_{1n} \\ a_{21} & & & & & \cdot \\ \cdot & & & & & \cdot \\ \cdot & & & & & \cdot \\ \cdot & & & & & \cdot \\ a_{n1} & & \cdot & \cdot & \cdot & a_{nn} \end{bmatrix}, \quad B = (b_1, ..., b_n).$$

Then the quadratic form is

$$q(x_1, ..., x_n) = X^t A X, \tag{5.45}$$

and

$$f(x_1, ..., x_n) = X^t A X + B X + c. \tag{5.46}$$

Then by a suitable orthogonal transformation P, the quadric is carried to Equation (5.39), where PAP^t is diagonal. Once A is diagonal, linear terms are eliminated by the translation (5.40) or else (5.42) is used.

REMARK 5.13. *In proving the following theorem we apply the same kind of algebraic rearrangements of the terms of the given quadratic equation with real coefficients, for instance, applying the so-called method of* ***'completing the square'*** *and* ***'linearly changing the variables'*** *whenever they are needed, etc., as we have done in the proof of Theorem 5.10 above. We note that these two algebraic rearrangement of the terms of a polynomial equation with real coefficients is always possible as long as the degree of the polynomial is two.*

But, for a general polynomial equation in several variables of more degree than 2 with real coefficients, these methods are not sufficient. For example, when the degree of a real polynomial in several coefficients is 3, we are going to have some more difficulty but it's not going to be impossible. For more about this, refer to Section 3. Cubic Equations, Chapter 6. Discriminants and Galois Groups in my book, Computational Algebra,. *There, you will see the additional method of* ***'completing the cube'*** *to these two.*

Here is the classification in the case of having three variables:

THEOREM 5.11. *The congruence classes of nondegenerate quadrics in* $\mathbb{R}^3$ *are congruent to one of the followings: For* $a_{11}, a_{22}, a_{33} > 0$,

- **a:** *(Ellipsoids)* $a_{11}x_1^2 + a_{22}x_2^2 + a_{33}x_3^2 - 1 = 0$,
- **b:** *(1-sheeted hyperboloids)* $a_{11}x_1^2 + a_{22}x_2^2 - a_{33}x_3^2 - 1 = 0$,
- **c:** *(2-sheeted hyperboloids)* $a_{11}x_1^2 - a_{22}x_2^2 - a_{33}x_3^2 - 1 = 0$,
- **d:** *Elliptic paraboloids)* $a_{11}x_1^2 + a_{22}x_2^2 - x_3 = 0$,
- **e:** *(Hyperbolic paraboloids)* $a_{11}x_1^2 - a_{22}x_2^2 - x_3 = 0$. □

REMARK 5.14. *If a quadratic equation* $f(x_1, x_2)$ *is given, since its associated quadratic form is symmetric, we can determine the type of conic it represents most easily by allowing non-orthogonal changes of coordinates.*

For example, if the associated quadratic form q is positive definite, then the conic is either ellipse, or else it is degenerate (a single point or empty). For, from the original equation in (5.35) $a_{11} > 0$ *and* $a_{11}a_{22} - a_{12}^2 > 0$ *imply* $a_{11} > 0$ *and* $a_{22} > 0$*; hence part b and part c of Theorem 5.10 does not happen. To distinguish these cases, arbitrary changes of coordinates are permissible. A non-orthogonal coordinate change will distort the conic, but it will not change an ellipse into a hyperbola or a degenerate conic.*

REMARK 5.15. *Here is the graph of an ellipsoid:*

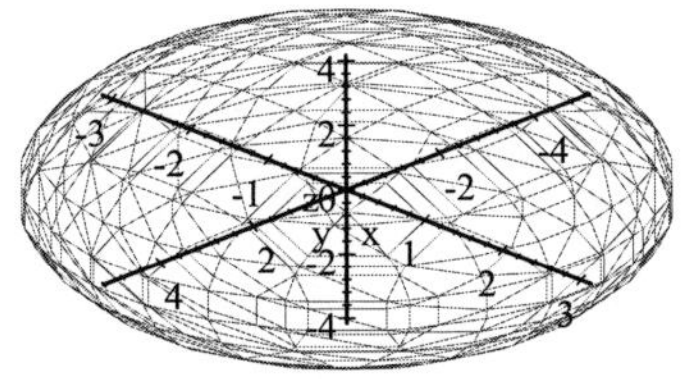

$$\frac{x^2}{3^2} + \frac{y^2}{5^2} + \frac{z^2}{4^2} = 1$$

REMARK 5.16. *Here is the graph of a 1-sheeted hyperboloid:*

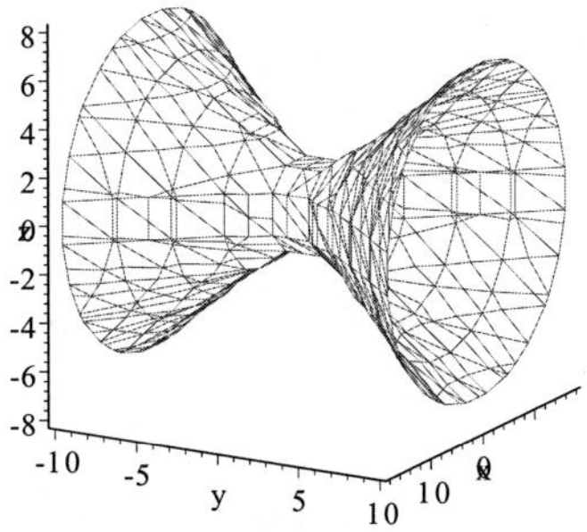

$$\frac{x^2}{5^2} - \frac{y^2}{3^2} + \frac{z^2}{2^2} = 1$$

REMARK 5.17. *Here is the graph of a 2-sheeted hyperboloid:*

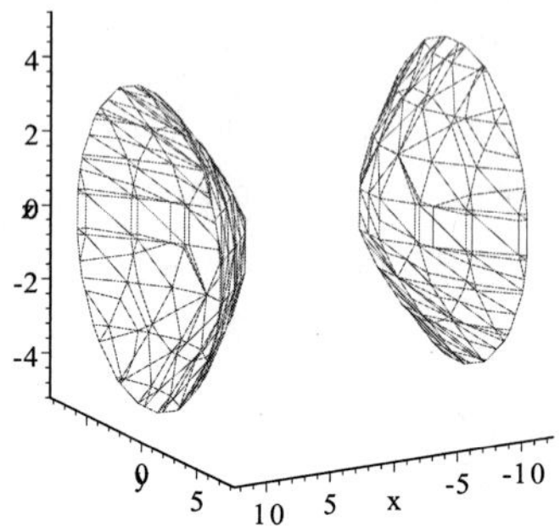

$$\frac{x^2}{5^2} - \frac{y^2}{3^2} - \frac{z^2}{2^2} = 1$$

REMARK 5.18. *Here is the graph of an elliptic paraboloid:*

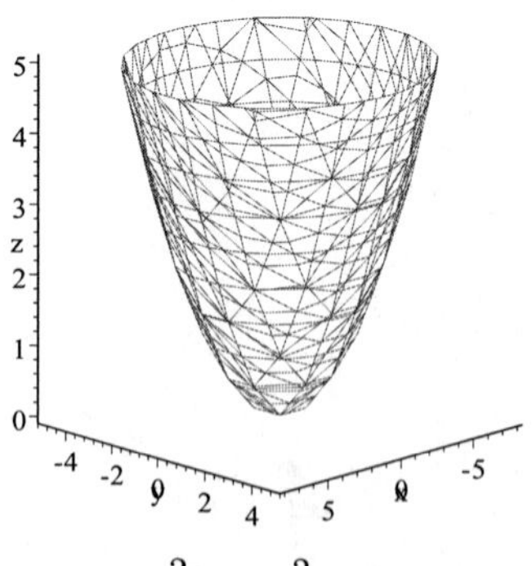

$$\frac{x^2}{5^2} + \frac{y^2}{3^2} = \frac{z}{2}$$

REMARK 5.19. *Here is the graph of a hyperbolic paraboloid:*

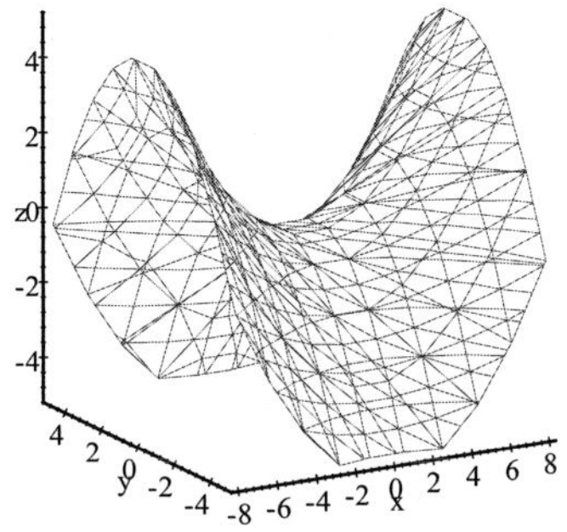

$$\frac{x^2}{5^2} - \frac{y^2}{3^2} = \frac{z}{2}$$

EXAMPLE 5.11. *Consider the locus*

$$x_1^2 + x_1x_2 + x_2^2 + 4x_1 + 3x_2 + 4 = 0. \tag{5.47}$$

The associated matrix is

$$A = \begin{bmatrix} 1 & \frac{1}{2} \\ \frac{1}{2} & 1 \end{bmatrix},$$

which is positive definite by Theorem (5.3). Since A is symmetric,we can diagonalize A by the non-orthogonal substitution $X' = PX$ for some invertible matrix P (which will be determined later below), i.e., by establishing congruence relations..Then, $X = P^{-1}X'$, and we want

$$X^tAX = X'^t(P^{-1})^tAP^{-1}X' \tag{5.48}$$

to be diagonal. By elementary row operation, we get

$$QA = \begin{bmatrix} 1 & 0 \\ -\frac{1}{2} & 1 \end{bmatrix} \begin{bmatrix} 1 & \frac{1}{2} \\ \frac{1}{2} & 1 \end{bmatrix} = \begin{bmatrix} 1 & \frac{1}{2} \\ 0 & \frac{3}{4} \end{bmatrix}.$$

Then since A is symmetric, we get

$$QAQ^t = \begin{bmatrix} 1 & 0 \\ 0 & \frac{3}{4} \end{bmatrix}.$$

Thus by (5.48), $Q = (P^{-1})^t$*; hence*

$$P = \begin{bmatrix} 1 & -\frac{1}{2} \\ 0 & 1 \end{bmatrix}^{-1} = \begin{bmatrix} 1 & \frac{1}{2} \\ 0 & 1 \end{bmatrix}.$$

Thus, we have

$$P = \begin{bmatrix} 1 & \frac{1}{2} \\ 0 & 1 \end{bmatrix}, \quad (P^{-1})^t A P^{-1} = \begin{bmatrix} 1 & 0 \\ 0 & \frac{3}{4} \end{bmatrix}, \quad BP^{-1} = (4, 1),$$

to obtain

$$x_1'^2 + \frac{3}{4} x_2'^2 + 4x_1' + x_2' + 4 = 0.$$

Completing the square yields

$$x_1''^2 + \frac{3}{4} x_2''^2 - \frac{1}{3} = 0,$$

which is an ellipse. Thus original locus of (5.47) represents an ellipse too. On the other hand, if we change the constant term of (5.47) to 5, the locus becomes empty.

Exercises for Section 5.7

EXERCISE 5.41. *Determine the type of the quadric*

$$x^2 + 4xy + 2xz + z^2 + 3x + z - 6 = 0.$$

EXERCISE 5.42. *Discuss all degenerate loci for conics.*

EXERCISE 5.43. *Describe the degenerate quadrics, that is, those which are not listed in Theorem 5.11.*

CHAPTER 6

Generators and Linear Relations

1. Modules and Determinants of R-Matrices

1.1. Introduction. When we study we have seen many examples of "linear relations". For example, a set $\{v_1, ..., v_n\}$ of vectors is called *linearly independent* if

$$c_1v_1 + ... + c_nv_n = 0,$$

for some scalars c_i's at least one of which is non-zero. Or, systems of linear equations, etc.

On the other hand, we recall that a vector space V over a field F is defined as a module over F. Then since every vector space, regardless of being finitely generated or not, has a basis, it is a free module of finite rank or infinite rank. (Remark: An official definition of a free module will be given in the next section.) As we shall see in the next section, this, namely, "every module over a field is free", is very special in the sense that modules M over a commutative ring R are not free in general.

In this chapter, we will systematically develop a theory for generators and linear relations among them up to a degree sufficient enough for us to study $F[t]$-modules, which will eventually lead us to better understand endomorphisms, i.e., linear operators, on a vector space in general. This will require us to study the structure of finitely generated abelian groups. We will need some knowledge on the polynomial ring $F[t]$, for instance, it is a *PID*, but it is going to be OK, because we are very familiar with the ring $\mathbb{Z}$ of integers.

We can also think of generators and non-linear relations among them, but they are better understood if we study them after learning the Theory of Groebner Bases first.

1.2. Modules. We first recall that a **module** M over a commutative ring R, or shortly, a R-module, is, by definition, an abelian group together with a scalar multiplication $R \times M \longrightarrow M$, sending $(r, v) \mapsto rv$, satisfying the following 4 axioms: for each $v, v' \in M$ and each scalar $r, s \in R$,

(1) $1v = v$,
(2) $(rs)v = r(sv)$,
(3) $(r+s)v = rv + sv$,
(4) $r(v+v') = rv + rv'$.

EXAMPLE 6.1. *If F is a field then an F-module is just a vector space over F.*

EXAMPLE 6.2. *Any commutative ring R is a module over itself. In this case, the above 4 axioms for the scalar multiplication for a module are simply those axioms defining it as a ring.*

The concepts of basis and linear independence can be carried over from vector spaces over a field K to modules over a commutative ring R without any change:

DEFINITION 6.1. *Let M be an R-module.*

a: *A set $\{v_1, ..., v_n\}$ of elements of M is said to* ***generate*** *(or* ***span****) M if each $v \in M$ is an R-linear combination:*

$$v = r_1 v_1 + ... + r_n v_n,$$

with $r_i \in R$. In that case, the v_i are called ***generators****.*

b: *M is said to be* ***finitely generated*** *if it has a finite set of generators.*

c: *We call a set of elements $\{v_1, ..., v_n\}$ of M (linearly)* ***independent*** *if the following condition holds:*

$$r_1 v_1 + ... + r_n v_n = 0 \text{ with } r_i \in R \quad \text{implies} \quad \text{each } r_i = 0 \text{ for } i = 1, ..., n$$

d: *Moreover, such a set is called a* ***basis*** *of M if it is a generating set of M.*

We then have an easy

LEMMA 6.1. *A subset* $\{v_1, ..., v_n\}$ *of an* R*-module* M *is a basis if and only if every element* v *of* M *is expressed as a linear combination*

$$v = r_1 v_1 + ... + r_n v_n$$

of the v_i*'s in a unique way, i.e., with unique coefficients* r_i*'s in* R. □

DEFINITION 6.2. (1) *A* ***homomorphism***, *or* ***linear map***, *of* R*-modules* $\varphi\colon M \longrightarrow N$ *is a map which is compatible with the laws of composition:*

$$\varphi(v + w) = \varphi(v) + \varphi(w) \qquad \text{and} \qquad \varphi(rv) = r\varphi(v),$$

for each $r \in R$ *and for each* $v, w \in M$.

(2) *A bijective homomorphism of* R*-modules is called an* ***isomorphism***.

DEFINITION 6.3. *A* ***submodule*** N *of a module* M *is a non-empty subset of* M *such that it itself is closed under addition and scalar multiplication on* M.

Like in the case of linear maps of vector spaces, we can easily prove that

LEMMA 6.2. *The kernel of a homomorphism* $\varphi\colon M \longrightarrow N$ *of* R*-modules is a submodule of* M*, and the image of* $\varphi\colon M \longrightarrow N$ *is a submodule of* N. □

Now, the most non-trivial but obvious examples are the modules R^n of R-vectors.

DEFINITION 6.4. *The row or column vectors in* R^n *with entries in the ring* R *are called* R*-**vectors**. The laws of composition for* R*-vectors are* *the same as for vectors with entries in a field F:*

$$\begin{bmatrix} a_1 \\ \cdot \\ \cdot \\ \cdot \\ a_n \end{bmatrix} + \begin{bmatrix} b_1 \\ \cdot \\ \cdot \\ \cdot \\ b_n \end{bmatrix} = \begin{bmatrix} a_1 + b_1 \\ \cdot \\ \cdot \\ \cdot \\ a_n + b_n \end{bmatrix} \quad \textit{and} \quad r \begin{bmatrix} a_1 \\ \cdot \\ \cdot \\ \cdot \\ a_n \end{bmatrix} = \begin{bmatrix} ra_1 \\ \cdot \\ \cdot \\ \cdot \\ ra_n \end{bmatrix}.$$

Certainly, R^n together with such laws of composition is an R-module. This module is very special because it is a set of elements

$$e_i = \begin{bmatrix} 0 \\ \cdot \\ 1 \\ \cdot \\ 0 \end{bmatrix},$$

where the i-th entry is 1 but all other entries are 0, for each $i = 0, ..., n$. They are called the n **unit elements** of R^n, or the **(standard) basis** of R^n. As the standard basis in the case of the vector space F^n where F is a field, every element $v = (a_1, ..., a_n)^t \in R^n$ is uniquely expressed as an R-linear combination:

$$v = a_1 e_1 + ... + a_n e_n$$

of the n unit vectors in R^n with a_i as coefficients. The R-module R^n with the standard basis $\{e_i\}$ satisfies a very special property:

THEOREM 6.1. *Given n elements $v_1, ..., v_n$ of an R-module M, the linear map t: $R^n \longrightarrow M$ defined by*

$$\sum_{i=1}^{n} a_i e_i \longmapsto \sum_{i=1}^{n} a_i v_i$$

is the only linear map from R^n to M that sends e_i to v_i for each $i = 1, ..., n$.

PROOF. One can easily check that sending any $v = \sum_{i=1}^n a_i e_i \longmapsto \sum_{i=1}^n a_i v_i$ is a linear map from R^n to M. Of course this map sends each e_i to v_i by definition. If T is a linear map: $R^n \longrightarrow M$ which sends each

e_i to v_i for each $i = 1, ..., n$, then it certainly sends linear combinations to linear combinations: $\sum_{i=1}^{n} a_i e_i \longmapsto \sum_{i=1}^{n} a_i v_i$; hence $t = T$. □

Thus, R^n is a free R-module, according to the following definition.

DEFINITION 6.5. *Let X be a subset of an R-module F and $j\colon X \hookrightarrow F$ be the insertion of X in F. Then F is called* ***a free module on*** *X, or* ***free*** *on X,* ***(and X a set of free generators*** *for the module F) if we have the following universally repelling property: whenever we have a map $f\colon X \longrightarrow M$ to an R-module M, there exists a unique (repelling) linear map $t\colon F \longrightarrow M$ such that $f = t \circ j$. In this case, any set X of free generators is also called a* ***basis*** *for the free module F.*

Thus, the above theorem says

THEOREM 6.2. *The R-module R^n is a free module on the set $\{e_1, ..., e_n\}$ of the standard basis of R^n.* □

COROLLARY 6.1. *An R-module with a basis $\{v_1, ..., v_n\}$ is isomorphic to R^n.* □

DEFINITION 6.6. *Such an R-module is also called a* ***free*** *R-module.*

REMARK 6.1. *Thus, we immediately see that any surjective linear map onto a finitely generated R-module is completely determined by the images of free generators of a free R-module, and this observation is going to be the starting point of the main issue of this chapter, generators and linear relations among them. See one of the next section titles.*

We are now going to see that there are plenty of R-modules which are not free.

LEMMA 6.3. *Abelian groups and $\mathbb{Z}$-modules are equivalent concepts.*

PROOF. By definition, a $\mathbb{Z}$-module is an abelian group. Conversely, any abelian group V can be made into a $\mathbb{Z}$-module in exactly one way by the rules:

$$\begin{aligned} nv &= v + ... + v = \text{“}n \text{ times } v\text{”}, \\ (-n)v &= -(nv), \end{aligned}$$

for any positive integer n. Then those 4 axioms for a module is automatically satisfied. □

COROLLARY 6.2. *The ring $\mathbb{Z}$ of integers is a $\mathbb{Z}$-module.*

EXAMPLE 6.3. *Since the set $\mathbb{Z}$ consists of infinitely many elements, any finite abelian group cannot be possibly isomorphic to a free $\mathbb{Z}$-module $\mathbb{Z}^n$ for some $n > 0$, unless it is the zero group; hence they are not free at all. A free $\mathbb{Z}$-module is also called a* ***free abelian group****.*

This example and Remark 6.1 certainly suggests that we need to extend the concept of quotient group to modules.

DEFINITION 6.7. *Let M be an R-module and N be its submodule. Then the set M/N of cosets $v + N$ for various $v \in M$, sometimes called the* ***quotient*** *of M by N, becomes an R-module under the following binary operations and scalar multiplication:*

$$\begin{aligned} (v+N)+(w+N) &= vw+N, \\ s.(v+N) &= sv+N, \end{aligned}$$

for each $v \in M$ and for each $s \in R$.

The best example one can easily understand quotient modules may be the following example:

EXAMPLE 6.4. *Let $6\colon\ \mathbb{Z} \longrightarrow \mathbb{Z}$ be "multiplication by 6". Then this map is clearly a homomorphism of $\mathbb{Z}$-modules, and hence its kernel $6\mathbb{Z}$ is a submodule of $\mathbb{Z}$. Thus, elements of the quotient module $\mathbb{Z}/6\mathbb{Z}$ are those remainders of integers v modulo 6. More precisely, by Euclid's algorithm we have a unique expression,*

$$v = 6q + r$$

with uniquely determined quotient q and the remainder r, $0 \leq r < 6$; by definition elements of $\mathbb{Z}/6\mathbb{Z}$ are $r+ 6\mathbb{Z}$.

We note that any two integers s, t with the same remainder r defines the same coset, $r + 6\mathbb{Z}$; that is, $s + 6\mathbb{Z} = t + 6\mathbb{Z}$, i.e., they, s and t, *"represents" the same coset* $r + 6\mathbb{Z}$. *(They are called* ***representatives*** *of the coset they represent.) We also note in this case that* 6 *divides* $(s-t)$ *if and only if* $(s-t)+6\mathbb{Z} = 0+6\mathbb{Z}$, *i.e.,* $s-t \in 6\mathbb{Z}$.

Following ideas described in this example, one can easily prove

PROPOSITION 6.1. *Then we have:*

(1) *The two rules in the above definition are* ***well-defined****. That is, they are independent of representatives of cosets they represent; and consequently* M/N *under these rules become an* R*-module, called the quotient module of* M *by* N.
(2) *The canonical map* $\pi\colon M \longrightarrow M/N$ *which sends* $v \mapsto \bar{v} := v + N$ *is a homomorphism of* R*-modules which is surjective with kernel* N.
(3) *If* $f\colon M \longrightarrow M'$ *is any homomorphism of* R*-modules whose kernel contains* N, *then there exists a unique homomorphism*

$$\bar{f} : M/N \longrightarrow M'$$

such that the involved diagram is commutative, that is, $f = \bar{f} \circ \pi$.
(4) *Moreover, if kernel of* f *is the same as* N, *then* $\bar{f}$ *gives an isomorphism of* M/N *with its image,* $f(M)$.
(5) *There exists a one-to-one and onto correspondence between submodules* $\bar{S}$ *of* M/N *and those* S *of* M *which contains* N, *defined by*

$$S = \pi^{-1}(\bar{S}) \qquad \text{and} \qquad \bar{S} = \pi(S).$$

Moreover, if S *and* $\bar{S}$ *are corresponding submodules, then we have an isomorphism of* R*-modules:*

$$V/S \xrightarrow{\sim} (M/N)/\pi(S).$$

PROOF. We leave the proof of this proposition as an exercise. □

1.3. Determinants. Matrices with entries in a commutative ring R can be manipulated in the same way as matrices with entries in a field F. This is because the operations of matrix addition and matrix multiplication are defined as in the case of matrices with entries in a field F; consequently they satisfy similar rules.

DEFINITION 6.8. *A matrix with entries in a commutative ring R is called an R-**matrix**.*

The determinant of an $n \times n$ R-matrix $A = (a_{ij})$ can be computed by any of the rules for matrices with entries in a field F. In particular, it is convenient to use the complete expansion,

$$\det(A) = \sum_p sgn(p) \cdot a_{1p(1)} \cdots a_{np(n)},$$

where the sum is taken over all permutations, i.e., bijections, of the set $\{1, 2, ..., n\}$, because it exhibits the determinant as a polynomial in the n^2 matrix entries a_{ij}.

The usual rules for determinant with entries in a field F apply, in particular

$$\det(AB) = \det(A)\det(B).$$

COROLLARY 6.3. *The invertible $n \times n$ matrices A with entries in a commutative ring R are those matrices whose determinants are units in R.*

PROOF. If A has a multiplicative inverse A^{-1} with entries in R, then

$$\det(A)\det(A^{-1}) = \det(I) = 1,$$

and hence the determinant of an invertible R-matrix is a unit, i.e., an invertible element, in R.

Conversely, if A is an R-module whose determinant $\det(A)$ is a unit, then we can compute its inverse using Cramer's Rule:

$$\det(A)I = A \cdot adj(A),$$

where the adjoint matrix $adj(A)$ is calculated from A by taking determinants of minors. This rule also holds in any ring, so

$$A^{-1} = adj(A)/\det(A).$$

$\square$

REMARK 6.2. *The determinant of an invertible matrix of size* 1×1 *with entries in* $\mathbb{Z}$ *is either* $+1$ *or* -1*, because there are only two invertible elements in* $\mathbb{Z}$*. But, if* $n > 1$*, then there are always reasonably many matrices, because elementary matrices of the type*

$$mulrow(i,u) = \begin{bmatrix} 1 & & & & & & \\ & \ddots & & & & & \\ & & 1 & & & & \\ & & & u & & & \\ & & & & 1 & & \\ & & & & & \ddots & \\ & & & & & & 1 \end{bmatrix},$$

where u *is a unit in* R*, are invertible, and elementary matrices of the type*

$$addrow(j,i,a) = \begin{bmatrix} 1 & & & & & & \\ & \ddots & & & & & \\ & & 1 & & & & \\ & & & \ddots & & & \\ & & & & \ddots & & \\ & & a & & & \ddots & \\ & & & & & & 1 \end{bmatrix} \tag{6.1}$$

$$\text{or} \quad \begin{bmatrix} 1 & & & & & & \\ & \ddots & & & a & & \\ & & \ddots & & & & \\ & & & \ddots & & & \\ & & & & 1 & & \\ & & & & & \ddots & \\ & & & & & & 1 \end{bmatrix} \tag{6.2}$$

where $i \neq j$ *and* $a \in R$ *have determinant* 1.

1.4. The matrices of homomorphisms of free R-modules. Let R^n be a free R-module and $\mathbf{B} = (v_1, ..., v_n)$ an ordered subset of an R-module M, and φ: $R^n \longrightarrow M$ be a map, defined by

$$X = \begin{bmatrix} x_1 \\ \cdot \\ \cdot \\ \cdot \\ x_n \end{bmatrix} \mapsto \mathbf{B}X = v_1x_1 + ... + v_nx_n.$$

Then one can easily check that

LEMMA 6.4. (1) *it is a homomorphism of R-modules, and*

(2) *this homomorphism is surjective if and only the set $\mathbf{B} = (v_1, ..., v_n)$ generates M, and*

(3) *the homomorphism is injective if and only if the set $\mathbf{B} = (v_1, ..., v_n)$ is (linearly) independent.* □

Thus, the homomorphism is an isomorphism if and only if $\mathbf{B} = (v_1, ..., v_n)$ is a basis of M , in which case M is a free module. Thus, a finitely generated R-module M has a basis if and only if it is free. As we have seen before, most finitely generated modules have no bases.

Computation with bases of free R-modules can be done in much the same way as with bases of vector spaces, using matrices with entries in R. In particular, we can speak of the **coordinate vector** $X = (x_1, ..., x_n)^t$ in R^n of an element $v \in M$, with respect to a basis $\mathbf{B} = (v_1, ..., v_n)$ of M. It is the unique column vector $X \in R^n$ such that

$$v = \mathbf{B}X = v_1x_1 + ... + v_nx_n.$$

Now, if two bases $\mathbf{B} = (v_1, ..., v_n)$ (old basis) and $\mathbf{B}' = (v'_1, ..., v'_r)$ (new basis) for the same free module M are given, then the **matrix of change of basis** is obtained as usual by writing the elements v_j of the old basis $\mathbf{B}$ as linear combination of the new basis:

$$\mathbf{B} = \mathbf{B}'P,$$

or,

$$v_j = \sum_{j=1}^{r} v_i' p_{ij}.$$

As with vector spaces, we have

PROPOSITION 6.2. *Any two bases of the same free finitely generated module M over a non-zero commutative ring R have the same cardinality.*

PROOF. Let $\mathbf{B}$ and $\mathbf{B}'$ be two bases of a free R-module M. We write elements of $\mathbf{B}$ as a linear combination of elements of $\mathbf{B}'$ and we write element of $\mathbf{B}'$ as a linear combination of elements of $\mathbf{B}$:

$$\mathbf{B} = \mathbf{B}'P \qquad \text{and} \qquad \mathbf{B}' = \mathbf{B}Q.$$

Thus, we have

$$\mathbf{B} = \mathbf{B}QP.$$

Then since $\mathbf{B}$ is a basis, there exists only one way to write each v_j as a linear combination of the v_j's:

$$v_j = 1v_j,$$

or $\mathbf{B} = \mathbf{B}I$; consequently we see that $QP = I$, and similarly $PQ = I$ holds.

We now note that P is an $r \times n$ matrix and Q is an $n \times r$ matrix, so we need to prove that $r = n$, in order to conclude that P and Q are indeed invertible R-matrices. For this, we suppose that $r > n$. Then we can always make P and Q square matrices by adding enough zeros:

$$I = PQ = \begin{bmatrix} P & 0 \end{bmatrix} \begin{bmatrix} Q \\ 0 \end{bmatrix}.$$

But then the determinant of the product PQ is equal to 1, whereas the product of the determinants, 0's, of these new square matrices with enough zeros is zero, which is impossible. Thus, we must always have $r = n$. □

DEFINITION 6.9. *The number of elements of a basis for a finitely generated free R-module M is call the **rank** of M, instead of the dimension.*

As in the case of vector spaces, we have

COROLLARY 6.4. *Every homomorphism $\varphi\colon R^n \longrightarrow R^m$ of free R-modules is left multiplication by an R-matrix.* □

DEFINITION 6.10. *If $\varphi\colon M \longrightarrow N$ is a homomorphism of free R-modules with bases $\mathbf{B} = (v_1, ..., v_n)$ for M and $\mathbf{C} = (w_1, ..., w_m)$ for N, then the* ***matrix of the homomorphism*** *φ with respect to $\mathbf{B}$ and $\mathbf{C}$ is defined to be $A = (a_{ij})$, where*

$$\varphi(v_j) = \sum_{i=1}^{m} w_i a_{ij}, \qquad j = 1, 2, ..., n.$$

REMARK 6.3. *Thus, as in the case of linear maps for vector spaces, if $\mathbf{B} = (v_1, ..., v_n)$ is the old basis of M, $\mathbf{B}' = (v_1', ..., v_n')$ is a new basis of M, and $\mathbf{C} = (w_1, ..., w_m)$ is the old basis of N, $\mathbf{C}' = (w_1', ..., w_m')$ is a new basis of N, then the change of basis from the old basis to the new basis for M and N, respectively, gives us a new matrix A' of the homomorphism φ with respect to $\mathbf{B}'$ and $\mathbf{C}'$:*

$$A' = QAP^{-1},$$

where invertible R-matrices P and Q are determined by the relations

$$\mathbf{B} = \mathbf{B}'P \qquad \text{and} \qquad \mathbf{C} = \mathbf{C}'Q.$$

Exercises for Section 6.1

EXERCISE 6.1. *Verify that any commutative ring R is a module over itself.*

EXERCISE 6.2. *Let R be a ring, considered as an R-module. How many module homomorphisms $\varphi\colon R \longrightarrow R$ we can have?*

EXERCISE 6.3. *Let $\varphi\colon V \longrightarrow W$ be a homomorphism of modules over a (commutative) ring R, and let V' and W' be submodules of V and W, respectively. Prove that $\varphi(V')$ is a submodule of W and $\varphi^{-1}(W')$ is a submodule of V.*

EXERCISE 6.4. *An R-module is called **simple** if it has no non-trivial proper submodule.*

(1) *Prove that any simple module if isomorphic to R/M, where M is a maximal ideal of R.*
(2) *Prove Schur's Lemma: Let $\varphi\colon S \longrightarrow S'$ be a homomorphism of simple modules. Then either φ is the zero map or else it is an isomorphism.*

EXERCISE 6.5. *Let R be a ring, V an R-module, and let $End(V)$ the set of all endomorphisms of V. Prove that $End(V)$ is a non-commutative ring under composition of module homomorphisms as multiplication and addition defined by $(\varphi + \phi)(m) = \varphi(m) + \phi(m)$.*

EXERCISE 6.6. *Prove that $End(V)$ of simple module V is a field.*

EXERCISE 6.7. *The annihilator of an R-module V is by definition the subset I of R:*

$$I = \{r \in R : rv = 0 \text{ for each } v \in V\}.$$

(1) *Prove that I is an ideal of R.*
(2) *What is the annihilator of the $\mathbb{Z}$-module $\mathbb{Z}/2\mathbb{Z} \times \mathbb{Z}/3\mathbb{Z} \times \mathbb{Z}/4\mathbb{Z}$?*

EXERCISE 6.8. *Let I be an ideal of a ring R. Prove that I is a free R-module if and only if it is a principal ideal, generated by an element α which is not a zero-divisor in R.*

EXERCISE 6.9. *Prove that every subgroup of $\mathbb{Z}$ is a free $\mathbb{Z}$-module.*

EXERCISE 6.10. *Let A be the matrix of a homomorphism $\varphi\colon \mathbb{Z}^n \longrightarrow \mathbb{Z}^m$ between free modules.*

(1) *Prove that φ is injective if and only if the rank of A is n.*
(2) *Prove that φ is surjective if and only if the greatest common divisor of the determinants of the $m \times m$ minors of A is 1.*

2. Diagonalization of Integer Matrices

The main purpose of this section is about how to simplify an $m \times n$ integer matrix $A = (a_{ij})$ by a succession of elementary operations, using possibly elementary row operations and column operations. The same method will work for matrices with entries in a Euclidean domain (ED) and, with some modification, for matrices with entries in a principal ideal domain (PID). We need this procedure to classify abelian groups in a later section of this chapter.

DEFINITION 6.11. *An integral domain R is called a **Euclidean domain** if there exists a size function $\sigma\colon R - \{0\} \longrightarrow \mathbb{N} \cup \{0\}$ such that the division algorithm holds:*

(Division Algorithm) *Let $a \neq 0, b \in R$. Then there exist elements $q, r \in R$ such that $b = aq + r$ with either $r = 0$ or $\sigma(r) < \sigma(a)$.*

REMARK 6.4. *We do not require the elements q, r to be uniquely determined by a and b.*

EXAMPLE 6.5.
(1) $a \longmapsto |a|$ *(the absolute value of a) is the size function: $\mathbb{Z} - \{0\} \longrightarrow \mathbb{N} \cup \{0\}$.*
(2) $f(x) \longmapsto \deg(f(x))$ *(the degree of $f(x)$) is the size function: $F[x] - \{0\} \longrightarrow \mathbb{N} \cup \{0\}$.*
(3) $a \longmapsto |a|^2$ *(the square of the absolute value of a) is the size function: $\mathbb{Z}[i] - \{0\} \longrightarrow \mathbb{N} \cup \{0\}$, where $i = \sqrt{-1}$.*

We are going to apply to an integer matrix the following elementary operations:

(1) Add an integer multiple of one row to another, or add an integer multiple of one column to another;
(2) interchange two rows or two columns;
(3) multiply a row or a column by a unit u.

REMARK 6.5. *In contrast to those elementary row operations and column operations for matrices with entries in a field F, the scalar u in the* *third operation must be a unit, i.e., an invertible element of R, otherwise, the third operation is not going to have a corresponding inverse operation.*

THEOREM 6.3. *Let A be an $m \times n$ integer matrix. Then there exist products Q, P of elementary integer matrices as above, so that $A' = QAP^{-1}$ is a diagonal matrix:*

$$D = \begin{bmatrix} \begin{bmatrix} d_1 & & \\ & \ddots & \\ & & d_r \end{bmatrix} & \\ & 0 \end{bmatrix}, \tag{6.3}$$

where the diagonal entries d_i are non-negative and $d_1|d_2, d_2|d_3, \ldots$, that is, each diagonal entry d_i divides d_{i+1}.

REMARK 6.6. *As the following examples will show, Q and P in the theorem are not uniquely determined.*

PROOF. Our strategy is to perform a sequence of operations until we end up with a matrix:

$$\begin{bmatrix} d_1 & 0 \\ 0 & \begin{bmatrix} & \\ & B & \\ & \end{bmatrix} \end{bmatrix} \tag{6.4}$$

where d_1 divides each entry of B. When this is done, we work on B, inductively. The process is based on repeated division with remainder. We may assume that $A \neq 0$.

(Step 1) By permuting rows and columns, we move a non-zero entry with smallest absolute value to the upper left corner. Multiply the first row by -1 if necessary, so that this upper left entry a_{11} becomes positive.

We now try to clear out the first row and column. Whenever an operation produces a non-zero entry in the matrix whose absolute value is smaller than a_{11}, we go back to (Step 1) above and start the whole process over, and this process guarantees to reduce the size of a_{11} each time; hence we will not have to return (Step 1) infinitely often.

(Step 2) We choose a non-zero entry a_{i1} in the first column, with $i > 1$, and divide by a_{11}:

$$a_{i1} = a_{11}q + r,$$

where $0 \leq r < a_{11}$. Thus, we subtract q times (row 1) from (row i) in order to change a_{i1} with r.

If $r \neq 0$, then we go bake to (Step 1).

Otherwise, i.e., if $r = 0$, then we have produced a zero in the first column.

Finitely many repetitions of (Step 1) and (Step 2) result in a matrix in which $a_{i1} = 0$ for each $i > 1$.

Similarly, we may use the analogue of (Step 2) for column operations to clear out the first row, eventually ending up with a matrix in which the only non-zero entry in the first row and column is a_{11}, as required. However, a_{11} may not yet divide each entry of the matrix B.

(Step 3) Assume that a_{11} is the only non-zero entry in the first row and column, but that some entry b of B is not divisible by a_{11}. Then add the column of A which contains b to the first column to have an entry b in the first column.

We then go back to (Step 2). Division with remainder will now produce a smaller matrix entry, sending us back to (Step 1).

A finite sequence of these steps will produce a matrix of the form (6.4, which then allows us to proceed by induction. □

EXAMPLE 6.6. *Let*

$$A = \begin{bmatrix} 2 & -1 \\ 1 & 2 \end{bmatrix}.$$

We try to reduce A in two different ways:

$$\begin{bmatrix} 2 & -1 \\ 1 & 2 \end{bmatrix} \overset{\text{row}}{\longrightarrow} \begin{bmatrix} 1 & 2 \\ 2 & -1 \end{bmatrix} \overset{\text{row}}{\longrightarrow} \begin{bmatrix} 1 & 2 \\ 0 & -5 \end{bmatrix} \overset{\text{col}}{\longrightarrow} \begin{bmatrix} 1 & 0 \\ 0 & -5 \end{bmatrix} \overset{\text{row}}{\longrightarrow} \begin{bmatrix} 1 & 0 \\ 0 & 5 \end{bmatrix},$$

and hence $P^{-1} = \begin{bmatrix} 1 & -2 \\ 0 & 1 \end{bmatrix}$ *and* $Q = \begin{bmatrix} 1 & 0 \\ 0 & -1 \end{bmatrix} \begin{bmatrix} 1 & 0 \\ -2 & 1 \end{bmatrix} \begin{bmatrix} 0 & 1 \\ 1 & 0 \end{bmatrix}$,

so that

$$A' = QAP^{-1} = \begin{bmatrix} 1 & 0 \\ 0 & 5 \end{bmatrix}.$$

Now secondly,

$$\begin{bmatrix} 2 & -1 \\ 1 & 2 \end{bmatrix} \overset{\text{col}}{\longrightarrow} \begin{bmatrix} 1 & -1 \\ 3 & 2 \end{bmatrix} \overset{\text{col}}{\longrightarrow} \begin{bmatrix} 1 & 0 \\ 3 & 5 \end{bmatrix} \overset{\text{row}}{\longrightarrow} \begin{bmatrix} 1 & 0 \\ 0 & 5 \end{bmatrix},$$

and hence $Q = \begin{bmatrix} 1 & 0 \\ -3 & 1 \end{bmatrix}$, $P^{-1} = \begin{bmatrix} 1 & 0 \\ 1 & 1 \end{bmatrix} \begin{bmatrix} 1 & 1 \\ 0 & 1 \end{bmatrix}$, *so that*

$$A' = QAP^{-1} = \begin{bmatrix} 1 & 0 \\ 0 & 5 \end{bmatrix}.$$

Thus, we do not need to follow the above systematic method to obtain A'.

We note that the key ingredient in the above proof is the *division algorithm*, and hence the same proof will work when $\mathbb{Z}$ is replaced with any Euclidean domain.

THEOREM 6.4. *Let R be a Euclidean domain, for example a polynomial ring $F[t]$ in one variable over a field F. Let A be an $m \times n$ matrix with entries in R. Then there exist products Q, P of elementary R-matrices such that $A' = QAP^{-1}$ is a diagonal matrix and each diagonal entry of A' divides the next, i.e., $d_1|d_2|d_3|...$, and each d_j is non-negative. If $R = F[t]$, then we can moreover normalize by requiring the polynomials d_i to be monic.* □

EXAMPLE 6.7. *Here is a diagonalization of a matrix with entries consisting of integer polynomials. Let*

$$A = \begin{bmatrix} (t-2)(t-1) & t-2 \\ (t-1)^3 & (t-2)(t-1) \end{bmatrix}.$$

This example will tell us something. If one attempts to reduce entries to zeros too quickly like:

$$A \overset{\text{col}}{\to} \begin{bmatrix} 0 & t-2 \\ (t-1)^2 & (t-2)(t-1) \end{bmatrix} \overset{\text{row}}{\to} \begin{bmatrix} 0 & t-2 \\ (t-1)^2 & 0 \end{bmatrix}$$
$$\overset{\text{col}}{\to} \begin{bmatrix} t-2 & 0 \\ 0 & (t-1)^2 \end{bmatrix},$$

but $t-2$ *never divides* $(t-1)^2$*; hence the final matrix is not what we are looking for. The point we missed in this computation is that we divided out other entries with* $t-2$ *without first checking that* $t-2$ *is the smallest one we should have during our reduction process on the entries of* A.

In the following computation, we will see that $1 = \gcd(t-1, t-2)$*, the* greatest common divisor *of* $t-1$ *and* $t-2$*, is in fact the smallest one. Since* $(t-1)^3 = t^3 - 3t^2 + 3t - 1 = t \cdot (t-2)(t-1) + (t-1)$ *by the division algorithm,*

$$A \overset{\text{row}}{\to} \begin{bmatrix} (t-2)(t-1) & t-2 \\ t-1 & -(t-2) \end{bmatrix} \overset{\text{row}}{\to} \begin{bmatrix} (t-1) & -(t-2) \\ (t-2)(t-1) & t-2 \end{bmatrix}$$
$$\overset{\text{col}}{\to} \begin{bmatrix} 1 & -(t-2) \\ (t-2)t & t-2 \end{bmatrix} \overset{\text{row}}{\to} \begin{bmatrix} 1 & -(t-2) \\ 0 & (t-2)(t-1)^2 \end{bmatrix}$$
$$\overset{\text{col}}{\to} \begin{bmatrix} 1 & 0 \\ 0 & (t-2)(t-1)^2 \end{bmatrix},$$

Thus, $A' = \begin{bmatrix} 1 & 0 \\ 0 & (t-2)(t-1)^2 \end{bmatrix}$.

The diagonalization of integer matrices can be used to describe homomorphisms between free abelian groups, i.e., free $\mathbb{Z}$-modules.

THEOREM 6.5. *Let* $\varphi\colon M \longrightarrow N$ *be a homomorphism of free abelian groups. Then there exist bases of* M *and* N *such that the matrix of the homomorphism* φ *with respect to these bases is of the diagonal form in Theorem 6.3.*

PROOF. As we have already described in Remark 6.3, a homomorphism $\varphi\colon M \longrightarrow N$ of free abelian groups is described by a matrix A once bases for M and N are chosen. Then by Theorem 6.3 there exists a change of basis by invertible integer matrices P, Q which changes A to a diagonal matrix $A' = QAP^{-1}$. $\square$

Let $\varphi\colon M \longrightarrow N$ be a homomorphism of R-modules. Then we know that the kernel of φ, $\ker(\varphi)$, and the image of φ, $img(\varphi)$, are submodules of M and N, respectively.

REMARK 6.7. *Our general concern is that under what conditions they are finitely generated if both M and N are finitely generated R-modules. In particular, in the next section we are going to prove that every subgroup, i.e., $\mathbb{Z}$-submodule, of a finitely generated abelian group, i.e., $\mathbb{Z}$-module, is finitely generated. (This is because the ring $\mathbb{Z}$ is a Noetherian ring, as we shall see in the next section. See Proposition 6.5.) Before we move into a more general setting, we can say something when both M and N are free R-modules of finite rank, respectively.*

Let $\varphi\colon \mathbb{Z}^n \longrightarrow \mathbb{Z}^m$ be left multiplication by an $m \times n$ integer matrix A. The kernel of φ is the subgroup of $\mathbb{Z}^n$ which consists of the system of linear equations:

$$AX = 0. \tag{6.5}$$

Then by the above theorem, we can diagonalize A, say to $A' = QAP^{-1}$, where Q, P are products of elementary integer matrices. To solve (6.5), we make the change of variable $X' = PX$ and solve the diagonal system

$$A'X' = QAP^{-1}X' = 0,$$

say, $d_1x_1' = 0, ..., d_nx_n' = 0$. If $d_i = 0$ then x_i' is arbitrary, and if $d_i \neq 0$, then x_i' must be zero. Here $X = (x_1, ..., x_n)^t$ and $X' = (x_1', ..., x_n')^t$, as usual. Thus, we have solved $QAX = 0$. Then since Q is invertible, both $QAX = 0$ and $AX = 0$ have the same solutions, $X = P^{-1}X'$ of the original system.

Now, what is the image of left multiplication by an $m \times n$ integer matrix $A = \varphi \colon \mathbb{Z}^n \longrightarrow \mathbb{Z}^m$? By definition, it consists of vectors $B \in \mathbb{Z}^m$ such that the system of linear equations $AX = B$ has an integer solution. Since the image of each element e_i of the standard basis $(e_1, ..., e_n)$ of the domain $\mathbb{Z}^n$ is mapped under A to

$$Ae_i = A_i$$

where $A = [A_1, ..., A_n]$, each $B = AX = A_1x_1 + ... + A_nx_n$ for some x_i's. That is,

COROLLARY 6.5. *The image of left multiplication by A is generated by the columns of A as a $\mathbb{Z}$-submodule of $\mathbb{Z}^m$.*

Having said these as warm-up, we now, conversely, assume that we are given a finite generating set $A_1, ..., A_n \in \mathbb{Z}^m$ of a subgroup S of a free abelian group $\mathbb{Z}^m$. Then each generator A_i is the image of the free generator, i.e., the basis element, e_i of $\mathbb{Z}^n$ under left multiplication by the matrix $A := [A_1, ..., A_n]$; hence S is just the image of this left multiplication by $A \colon \mathbb{Z}^n \longrightarrow \mathbb{Z}^m$.

DEFINITION 6.12. *Such a matrix A is said to* **represent** *the generating set $(A_1, ..., A_n)$.*

In fact, this construction is universal by Theorem 6.1, and this interpretation of S as an image of a homomorphism can tell the meaning of left and right multiplication to A by invertible integer matrices Q and P^{-1}:

(1) Left multiplication $QA = Q[A_1, ..., A_n]$ to A by Q corresponds to a change of basis in the module $\mathbb{Z}^m$, the range of the map, and its effect is to multiply each of the generators A_i by the same quantity Q: $QA = [QA_1, ..., QA_n]$.
(2) Right multiplication AP^{-1} to A by P^{-1} represents a change of basis in the domain $\mathbb{Z}^n$, and its effect is to change the generating set $[A_1, ..., A_n]$ of S by a column operation P^{-1}: $AP^{-1} = [A_1, ..., A_n]P^{-1}$.

EXAMPLE 6.8. *For example,* $[A_1, A_2]\begin{bmatrix} 1 & r \\ 0 & 1 \end{bmatrix} = [A_1, A_2 + rA_1]$.

Combining these observations with diagonalization results in the following theorem:

THEOREM 6.6. *Let S be a subgroup of a free abelian group M of rank m. Then there exists a basis $(w_1, ..., w_m)$ of M and a basis $(u_1, ..., u_n)$ of S with the following properties:*

(1) $n \le m$,
(2) *for each $j \le n$ there exists a positive integer d_j such that $u_j = d_j w_j$, and*
(3) $d_1 | d_2 | d_3 | ....$

PROOF. By Proposition 6.5 which we are going to prove later, S is finitely generated as a $\mathbb{Z}$-submodule of M. Thus, let $(w_1, ..., w_m)$ be a basis $\mathbf{B}$ of M and $(u_1, ..., u_n)$ be a set of generators for S, which gives us an $m \times n$ matrix A which represents S as above. Then by the diagonalization theorem (that is, either one of Theorems 6.3, 6.4, or 6.5, there exist invertible integral matrices P, Q such that

$$A' = QAP^{-1}$$

is a diagonal matrix with diagonal entries d_j's with $d_j | d_{j+1}$ for each j. This exactly means that the old basis of the domain is changed to a new basis by P and there exists a new basis $\mathbf{B}' = (w_1', ..., w_m')$ of M such that $\mathbf{B} = \mathbf{B}'Q$ and the old generating set $(u_1, ..., u_n)$ of S is changed to a new generating set $(u_1', ..., u_n')$ of S. Then since the diagonal matrix

$$A' = \begin{bmatrix} \begin{bmatrix} d_1 & & \\ & \ddots & \\ & & d_n \end{bmatrix} & 0 \\ 0 & 0 \end{bmatrix},$$

where $d_j|d_{j+1}$ for each $j = 1, ..., n$ and each d_j is non-negative, represents the new generating set $(u'_1, ..., u'_n)$ of S, we have that each

$$u'_j = d_j w'_j.$$

This is because, since u'_j is the image of the jth basis element e'_j of the new basis of the domain by Theorem 6.1 for instance, which is exactly the same as the unique linear combination of the basis elements w'_i's with coefficients in the jth column of the diagonal matrix A':

$$[u'_1, ..., u'_n] = [A'(e'_1), ..., A'(e'_n)] = (w'_1, ..., w'_m)A'.$$

We then drop the primes to obtain the basis and generating set required, and the third condition $d_1|d_2|d_3|...$ is established.

Now, if $n > m$ then since A' is diagonal as above, its jth columns are all zero for $j > m$; hence each $u_j = 0w_1 + ... + 0w_m = 0$ for $j > m$. Since the zero element is useless as a generator we just throw it out of the generating set $(u_1, ..., u_n)$. For the same reason, we may throw out a generator u_j whenever $d_j = 0$; consequently, all the d_j will be positive and we will have $n \leq m$.

Finally, we need to verify that if the basis and generating sets are chosen so that $d_j > 0$ and $n \leq m$, then $(u_1, ..., u_n)$ is a basis of S. But, this very easy: Since it generates S, we only need to prove that $(u_1, ..., u_n)$ is (linearly) independent. Assume that $r_1u_1 + ... + r_nu_n = 0$. Then since each $u_j = d_jw_j$ and since $(w_1, ..., w_m)$ is a basis, each $r_jd_j = 0$; hence each $r_j = 0$ since each $d_j > 0$. This finishes the proof of the whole theorem. □

REMARK 6.8. *We note that if S is the zero subgroup, then we will end up in the above proof throwing out all the generators. To avoid this, we must adopt the convention that the empty set generates the zero module, as with vector spaces. Or else, we need to make a special mention of this exceptional case in the statement of the theorem.*

COROLLARY 6.6. *Every subgroup of a free abelian group of rank m is free, and its rank is at most m.* □

Theorem 6.6 is quite explicit: Let S be the subgroup of $\mathbb{Z}^m$ generated by the columns of a matrix A, and suppose that $A' = QAP^{-1}$ is a diagonal matrix in either one of Theorems 6.3, 6.4, or 6.5. To display S in the form asserted in Theorem 6.6, we rewrite this equation in the form

$$Q^{-1}A' = AP^{-1}. \tag{6.6}$$

Then as we have already seen right before Theorem 6.6, the columns of the matrix AP^{-1} form our new set of generators for S. Then since

$$A' = \begin{bmatrix} \begin{bmatrix} d_1 & & \\ & \ddots & \\ & & d_n \end{bmatrix} & 0 \\ 0 & 0 \end{bmatrix}$$

is diagonal, (6.6) also tells us that the new generators are multiples of the columns of Q^{-1}. We change the basis of $\mathbb{Z}^m$ from the standard basis $(e_1, ..., e_m)$ to the basis consisting of the columns of Q^{-1}, because the matrix of this change of basis is Q. Then the new generators u'_j are d_j multiples of the new basis elements w'_j.

EXAMPLE 6.9. *Let S be the subgroup of $\mathbb{Z}^2$ generated by the two columns of the matrix of Example 6.6. By the computation we have done there, we have:*

$$\begin{aligned} Q^{-1}A' &= \begin{bmatrix} 1 & 0 \\ 3 & 1 \end{bmatrix}\begin{bmatrix} 1 & 0 \\ 0 & 5 \end{bmatrix} = \begin{bmatrix} 1 & 0 \\ 3 & 5 \end{bmatrix}, \\ AP^{-1} &= \begin{bmatrix} 2 & -1 \\ 1 & 2 \end{bmatrix}\begin{bmatrix} 1 & 1 \\ 1 & 2 \end{bmatrix} = \begin{bmatrix} 1 & 0 \\ 3 & 5 \end{bmatrix}. \end{aligned}$$

Thus, we have $[w'_1, w'_2]A' = [u_1, u_2]P^{-1} = [u'_1, u'_2]$. This means that the new basis $(w'_1, w'_2) = \left(\begin{bmatrix} 1 \\ 3 \end{bmatrix}, \begin{bmatrix} 0 \\ 1 \end{bmatrix}\right)$, and the new generators of S are

$$[u'_1, u'_2] = [u_1, u_2]P^{-1} = [w'_1, w'_2]A' = [w'_1, w'_2]\begin{bmatrix} 1 & 0 \\ 0 & 5 \end{bmatrix} = (w'_1, 5w'_2).$$

Exercises for Section 6.2

EXERCISE 6.11. *Find a basis for the following submodules of* $\mathbb{Z}^3$*:*

(1) *The module generated by* $(1,0,-1),(2,-3,1),(0,3,1),(3,1,5)$.

(2) *The module of solutions of the system of equations*

$$\begin{cases} x+2y+3z &= 0 \\ x+4y+9z &= 0 \end{cases}$$

EXERCISE 6.12. *Reduce each of the following integer matrices to diagonal form by integer row and column operations:*

(1) $\begin{bmatrix} 3 & 1 \\ -1 & 2 \end{bmatrix}$.

(2) $\begin{bmatrix} 1 & 2 & 3 \\ 4 & 5 & 6 \end{bmatrix}$.

(3) $\begin{bmatrix} 3 & 1 & -4 \\ 2 & -3 & 1 \\ -4 & 6 & -2 \end{bmatrix}$.

EXERCISE 6.13. *Let* $A = \begin{bmatrix} 4 & 7 & 2 \\ 2 & 4 & 6 \end{bmatrix}$.

(1) *Determine integer matrices* P^{-1} *and* Q *such that* QAP^{-1} *is a diagonal matrix.*

(2) *Determine all integer solutions to the system of equations* $AX = 0$.

3. Generators and Linear Relations

3.1. Presentation matrices. In the previous section, we have studied homomorphisms between free $\mathbb{Z}$-modules of finite rank and we have proved that every $\mathbb{Z}$-submodule of a free $\mathbb{Z}$-module M of rank m is again free of finite rank $\leq m$. (See Corollary 6.6). In this section, we turn our attention to modules which are *not* free. Then, in the next section we are going to prove Proposition 6.5 which was needed in the proof of Theorem 6.6.

REMARK 6.9. *The main concern of this section can be heuristically explained as follows: One of the main purposes of Linear Algebra is to learn how to solve a system of linear equations where two different quantities appear, the known ones and the unknown ones, and then we study related subjects, showing up in many different parts of Mathematics, pure and/or applied, and other parts of Sciences near Mathematics.*

Here, what we should note is that those quantities called known ones can be actually unknown to us and all really matters are the fact that we consider certain symbols to be known to us and other symbols not yet to be known to us. That is, we regard certain quantities to be known to us and other quantities not yet known to us, in order for us to easily approach or handle them when we try or attempt to mathematically understand the subject under consideration.

This clearly means that their role as "known" things or "unknown" things can be interchanged, and this approach is already utilized in the theory of invariants of the late 19th century. On the other hand, for Hardware called Computers and Software called Computer Algebra Systems (CAS), considering either of them as known or unknown is quite arbitrary.

Thus, we would like to study what's going on among the known ones, i.e., the coefficients or the column vectors consisting of them, when the symbols for the unknown ones are anything. That is, when a system of linear relations among generators is given to us, we forget about the generators themselves by regarding them simply as symbols, but we would like to study how the column vectors of coefficients are linearly related, and if so, we would like to figure out in detail the "smallest or linearly independent" generating set of column vectors which generates the whole column vectors of coefficients.

In the next section, we will then apply those results developed in this section and the diagonalization procedure of the previous section to matrices in order to study the structure of abelian groups.

We start this section with an example:

EXAMPLE 6.10. *We consider an abelian group or $\mathbb{Z}$-module M generated by three elements (v_1, v_2, v_3), satisfying the following relations:*

$$\begin{aligned} 3v_1 + 2v_2 + v_3 &= 0 \\ 8v_1 + 4v_2 + 2v_3 &= 0 \\ 7v_1 + 6v_2 + 2v_3 &= 0 \\ 9v_1 + 6v_2 + v_3 &= 0 \end{aligned} \tag{6.7}$$

We can rewrite this into matrix multiplication,

$$(v_1, v_2, v_3)A = (0, 0, 0, 0),$$

where

$$A = \begin{bmatrix} 3 & 8 & 7 & 9 \\ 2 & 4 & 6 & 6 \\ 1 & 2 & 2 & 1 \end{bmatrix} \tag{6.8}$$

is an 3×4 integer matrix whose columns consist of the integer coefficients of the relations (6.7), and hence it will define a linear homomorphism:

$$\mathbb{Z}^4 \longrightarrow \mathbb{Z}^3,$$

as is explained in the previous section, right after Theorem 6.5. We now need a few more terminologies:

DEFINITION 6.13. (1) *Equations of the form*

$$a_1 v_1 + \ldots + a_m v_m = 0, \qquad a_i \in R \tag{6.9}$$

are called **relations** *among the elements $(v_1, \ldots, v_m)$.*

(2) *The R-vector $(a_1, \ldots, a_m)^t$ which determine the relation is called a* **relation vector**.

(3) *A* **complete set of relations** *is a set of relation vectors such that every relation vector is a linear combination of relation vectors in this set.*

Thus, our concern is how to figure out a complete set of relations from the given matrix A, because a matrix such as (6.8) will not describe the module M completely unless its columns form a complete set of relations.

As we shall see it soon, it becomes much easier for us to understand the concept of a complete set of relations if we work with homomorphism of free modules rather than directly with the relations or the relation vectors: Let A be an $m \times n$ matrix with entries in a commutative ring R. As we have already seen it, left multiplication by A is a homomorphism of R-modules

$$\varphi\colon R^n \longrightarrow R^m. \tag{6.10}$$

We know that the kernel of φ and the image of φ are the two natural objects we usually consider whenever we are given a homomorphism

$$\varphi : M \longrightarrow M'$$

of R-modules, but there is one more natural object we need to consider, which is usually called the **cokernel** of φ and defined by the quotient module:

$$M'/(im\ \varphi). \tag{6.11}$$

REMARK 6.10. *In categorical terminologies, the kernel of φ is the "subobject" of φ and the cokernel of φ is its dual, "quotient object" of φ. And we always have these two objects in the category of R-modules.*

If we denote the image of left multiplication by A by AR^n as usual, the cokernel of (6.10) is

$$R^m/AR^n,$$

and this cokernel is usually said to **be represented by the matrix** A. More generally, we have

DEFINITION 6.14. *Any isomorphism*

$$\sigma\colon R^m/AR^n \xrightarrow{\sim} M \tag{6.12}$$

is called a ***representation*** *of an R-module M, and we say that the matrix A is a* ***presentation matrix*** *for M if there is such an isomorphism.*

EXAMPLE 6.11. *The cyclic group $\mathbb{Z}/5\mathbb{Z}$ is presented as a $\mathbb{Z}$-module by the 1×1 integer matrix $[5]$. On the other hand, consider the $\mathbb{Z}$-module*

M presented by the matrix

$$\begin{bmatrix} 2 & -1 \\ 1 & 2 \end{bmatrix}$$

is generated by two elements v_1 and v_2 with relations vectors $(2,1)^t$ and $(-1,2)^t$:

$$\begin{array}{rcl} 2v_1 + v_2 & = & 0 \\ -v_1 + 2v_2 & = & 0 \end{array}.$$

It seems that the module requires two generators, but it is not the case. In this simple case, we can easily solve the first equation to see that $v_2 = -2v_1$, which is a linear combination of v_1, and substituting this into the second equation we obtain $-5v_1 = 0$. This means that the module M is generated by a single generator v_1, with a single relation $5v_1 = 0$. Thus, M is isomorphic to $\mathbb{Z}/5\mathbb{Z}$; hence this 2×2 matrix also represents the cyclic group $\mathbb{Z}/5\mathbb{Z}$.

Here is a systematic and theoretical method of finding a presentation of a given R-module M: Let M be a finitely generated R-module and so let $(v_1, ..., v_m)$ be its generating set, an ordered set. Then by Theorem 6.1, we have a canonical surjective homomorphism of R-modules

$$f \colon R^m \longrightarrow M,$$

which sends the column vector $(x_1, ..., x_m)^t = \sum_{i=1}^m x_i e_i \mapsto \sum_{i+1}^m x_i v_i = v_1 x_1 + ... + v_m x_m$, that is, sending each of the generating elements, the standard basis elements, e_i of R^m to each of the generating elements v_i. If we denote its kernel by W, we have by the first isomorphism theorem in Proposition 6.1 that M is isomorphic to R^m/W.

If the submodule W of R^m is not finitely generated, then we cannot go further anymore.

But if W is again a finitely generated R-module like the module M, say by a generating set $(w_1, ..., w_n)$, an order set, then we can repeat the above procedure and we have another surjective homomorphism

$$g \colon R^n \longrightarrow W,$$

as before, and hence the composition of g with the obvious inclusion homomorphism i: $W \hookrightarrow R^m$ gives us a homomorphism of R-modules,

$$\varphi\colon R^n \longrightarrow R^m,$$

which sends each element e_i of the standard basis $(e_1, ..., e_n)$ of R^n to w_i, an element of the generating set $(w_1, ..., w_n)$ of W.

$$e_i \longmapsto w_i = \begin{bmatrix} a_{1i} \\ \vdots \\ a_{mi} \end{bmatrix}, \qquad i = 1, ..., n$$

Thus, φ is left multiplication by the $m \times n$ matrix A consisting of w_i's. This is because, for any column vector $(x_1, ..., x_n)^t$ of R^n, we have

$$\begin{aligned} \begin{bmatrix} x_1 \\ \vdots \\ x_n \end{bmatrix} = \textstyle\sum_{i=1}^n e_i x_i \longmapsto \sum_{i=1}^n w_i x_i \ &= w_1 x_i + ... + w_n x_n \\ &= (w_1, ..., w_n) \begin{bmatrix} x_1 \\ \vdots \\ x_n \end{bmatrix} \\ &= A \begin{bmatrix} x_1 \\ \vdots \\ x_n \end{bmatrix}. \end{aligned}$$

By our construction, W is clearly the image of φ: $R^n \longrightarrow R^m$, and hence it is AR^n. Thus,

$$R^m/AR^n = R^m/W \xrightarrow{\sim} M,$$

that is, the $m \times n$ matrix A is a presentation matrix for M, where m is the number of generators of the module M and n the number of generators of the kernel of the surjective module homomorphism $R^m \longrightarrow M$. Since the w_i's generate W, these columns $w_i = (a_{1i}, ..., a_{mi})^t$, $i = 1, ..., n$ form a complete set of relations among the generators $(v_1, ..., v_m)$ of the module M giving us

$$(v_1, ..., v_n)A = 0.$$

Thus a presentation matrix A for a finitely generated R-module M is determined by

(1) A set of generators for M,
(2) a complete set of relations among these generators,

provided that the kernel W in the above explanation is a finitely generated R-module.

REMARK 6.11. *As we shall see it later, Proposition 6.5 will provide us a sufficient condition for it, namely, R being a Noetherian ring.*

In particular, we note that $\mathbb{Z}$ is a Noetherian ring because it is a PID (Principal Ideal Domain).

Since any presentation matrix depends on the choices (1) and (2) above, other matrices can present the same module, or isomorphic modules. Here are some rules for manipulating a matrix A without changing the isomorphism class of the module it presents. *The striking point is that the size of presentation matrix can be reduced*, which we have already seen in Example 6.11.

PROPOSITION 6.3. *Let A be an $m \times n$ presentation matrix for an R-module M. The following matrices A' present the same module M:*

(1) *$A' = QAP^{-1}$ for some invertible R-matrix Q of size $m \times m$ and some invertible R-matrix P of size $n \times n$;*
(2) *A' is obtained by deleting a column of zeros from A;*
(3) *A' is obtained from A by deleting the ith row and j^{th} column when the jth column of A is e_i.*

PROOF. (1) The module R^m/AR^n is isomorphic to M. Then since the change of A to QAP^{-1} corresponds to a change of basis in R^m and R^n, the isomorphism class of the quotient module does not change.

(2) A column consisting of zero corresponds to the trivial relation, $0v_1 + ... + 0v_n = 0$, which can be omitted.

(3) Suppose that the jth column of the matrix A is e_i. Then the corresponding relation is $1v_i = 0$ in M, and hence v_i can be left out of

the generating set $(v_1, ..., v_n)$ of M. Doing this changes the matrix A by deleting the ith row and j^{th} column of A. □

Thus as the following example shows, we can simplify a matrix quite a lot by these rules.

EXAMPLE 6.12. *Let A be the integer matrix in our first example of this section, Example 6.10. We apply elementary row and column operations (without row exchanges) by (1) of the above proposition, and then we apply (2) or (3) of the above proposition to obtain A'.*

$$A = \begin{bmatrix} 3 & 8 & 7 & 9 \\ 2 & 4 & 6 & 6 \\ 1 & 2 & 2 & 1 \end{bmatrix} \to \begin{bmatrix} 0 & 2 & 1 & 6 \\ 0 & 0 & 2 & 4 \\ 1 & 2 & 2 & 1 \end{bmatrix} \mapsto \begin{bmatrix} 2 & 1 & 6 \\ 0 & 2 & 4 \end{bmatrix}$$

$$\mapsto \begin{bmatrix} 2 & 1 & 6 \\ -4 & 0 & -8 \end{bmatrix} \mapsto \begin{bmatrix} -4 & -8 \end{bmatrix} \mapsto [4].$$

Thus, A presents the abelian group $\mathbb{Z}/4\mathbb{Z}$.

By definition, an $m \times n$ matrix presents a module by means of m generators and n relations among m generators. But as we have seen from this example, the number of generators and the number of relations depend on choices we have made, and hence they are not uniquely determined by the given module M.

EXAMPLE 6.13. *The 2×1 matrix $\begin{bmatrix} 4 \\ 0 \end{bmatrix}$ presents a module an abelian group M by means of two generators (v_1, v_2) and one relation $4v_1 = 0$. That is, $(v_1, v_2)\begin{bmatrix} 4 \\ 0 \end{bmatrix} = 0$. Thus, v_2 can be any element and $4v_1 = 0$. Thus, the group which it presents is isomorphic to the product group $\mathbb{Z}/4\mathbb{Z} \times \mathbb{Z}$. Here, since $4 \neq 1$, it is not an e_1.*

On the other hand, the 1×2 matrix $\begin{bmatrix} 4 & 0 \end{bmatrix}$ presents a group with one generator v_1 and two relations, of which the second relation is trivial relation $0v_1 = 0$, thus $\begin{bmatrix} 4 & 0 \end{bmatrix}$ is reduced to $[4]$; hence it is isomorphic to $\mathbb{Z}/4\mathbb{Z}$.

In the next section, we discuss the problem of finite generation of the module of relations.

Exercises for Section 6.3

EXERCISE 6.14. *In each case of the following matrix A, identify the abelian group which has the given presentation matrix A:*

$$\begin{bmatrix} 2 \\ 1 \end{bmatrix}, \begin{bmatrix} 0 \\ 5 \end{bmatrix}, \begin{bmatrix} 2 & 0 & 0 \end{bmatrix}, \begin{bmatrix} 1 & 0 \\ 1 & 1 \end{bmatrix},$$

$$\begin{bmatrix} 1 & 0 \\ 0 & 1 \\ 0 & 0 \end{bmatrix}, \begin{bmatrix} 2 & 3 \\ 1 & 2 \end{bmatrix}, \begin{bmatrix} 2 & 4 \\ 1 & 2 \end{bmatrix}, \begin{bmatrix} 1 & 2 \\ 3 & 4 \end{bmatrix}$$

4. ACC or Noetherian Condition

In our proof of Theorem 6.6, we needed an additional result, Proposition 6.5 below. Namely, to finish the proof of Theorem 6.6 we must prove Proposition 6.5. For this, we start our discussion with equivalence of two statements:

PROPOSITION 6.4. *Let M be an R-module. The following statements are equivalent:*

- **a:** *Every submodule N of M is finitely generated;*
- **b:** *ACC(Ascending Chain Condition) There is no infinitely strictly increasing chain*

$$N_1 \subset N_2 \subset \cdots$$

of submodules N_i of M.

PROOF. Assume that M satisfies ACC, and let N be a submodule of M. If $N = \{0\}$, then N is generated by 0. If not, then N contains a non-zero element, say v_1. Let N_1 be the submodule of M generated by v_1. Then since $v_1 \in N$, N_1 is automatically a submodule of N. If $N_1 = N$, then we are done. Otherwise, there exists $v_2 \in N - N_1$, and let N_2 be the submodule generated by both v_1 and v_2. If $N_2 = N$, then we are done. Otherwise, there exists $v_3 \in N - N_2$, and let N_2 be the submodule generated by v_1, v_2, and v_3. If $N_3 = N$, then we are done. Otherwise, we repeat this procedure, so that

$$N_1 \subset N_2 \subset N_3 \subset \cdots,$$

but there should exist k such that $N_k = N$ by hypothesis. Thus, N is finitely generated.

Conversely, assume that every submodule of M is finitely generated, and let

$$N_1 \subset N_2 \subset \cdots$$

be an infinite increasing chain of submodules of M. Define U to be the union of these submodules N_i. Then one can easily check that U is a submodule of M. Thus, it is finitely generated by hypothesis. Let $u_1, ..., u_r$ be generators for U. Then each u_i is in one of the modules N_i, and since the chain is increasing, there should exist an integer i such that all the generators are in N_i; hence U is also contained in N_i, and this means that we have

$$U \subset N_i \subset N_{i+1} \subset U.$$

Thus, $U = N_i = N_{i+1} = \cdots$ and hence the chain is not strictly increasing. □

REMARK 6.12. *Here is some historical remark: It is known that Emmy Noether, a Jewish female mathematician whose father was a well-known Algebraic Geometer of* 19th *century, Max Noether, first recognized the importance of the ACC condition above, thus the condition is usually called a "**Noetherian Condition**" in order to celebrate her deep role in developing Modern Algebra of very axiomatic nature. In fact, she is called as the Mother of Modern Algebra, mainly because of her famous Lecture Note, published by Van der Waerden in 1930s.*

DEFINITION 6.15. *Any R-module M satisfying one of the equivalent statements of Proposition 6.4 is called a **Noetherian module**.*

LEMMA 6.5. *Let φ: $M \longrightarrow M'$ be a homomorphism of R-modules.*

(1) *If both the kernel and the image of* φ *are finitely generated modules, so is* M.

(2) *If* M *is finitely generated by a generating set* $\{v_1, ..., v_n\}$ *and if* φ *is* surjective, *then* M' *is finitely generated by the images* $\varphi(v_i)$, $i = 1, 2, ..., n$.

PROOF. Let $\{u_1, ..., u_k\}$ be a generating set for the kernel and let

$$\{w_1, ..., w_m\}$$

be a generating set for the image. Let v_i be the preimage of w_i, i.e., an element of M such that $\varphi(v_i) = w_i$. In what follows we prove that $\{u_1, ..., u_k, v_1, ..., v_m\}$ generates M: For this, let $v \in M$ be arbitrary. Then $\varphi(v)$ is a linear combination of $\{w_1, ..., w_m\}$, say $\varphi(v) = a_1 w_1 + ... + a_m w_m$. Now consider $v' = a_1 v_1 + ... + a_m v_m$. Since $\varphi(v') = \varphi(v)$, we see that $\varphi(v - v') = \varphi(v) - \varphi(v') = 0$; hence $v - v' \in \ker(\varphi)$. Thus, $v - v'$ is a linear combination of elements of $\{u_1, ..., u_k\}$, say $v - v' = b_1 u_1 + ... + b_k u_k$. Therefore,

$$\begin{aligned} v &= v' + b_1 u_1 + ... + b_k u_k \\ &= a_1 v_1 + ... + a_m v_m + b_1 u_1 + ... + b_k u_k. \end{aligned}$$

This shows that the set $\{u_1, ..., u_k, v_1, ..., v_m\}$ generates M, and this proves the first assertion.

The proof of the second assertion is very easy. □

COROLLARY 6.7. *Let* N *be a submodule of an* R*-module* M.

(1) *If both* N *and* M/N *are finitely generated, then so is* M.

(2) *If* M *is finitely generated, then so is* M/N. □

We recall that ideals in a commutative ring R, viewed as an R-module, are submodules, and vice versa.

DEFINITION 6.16. *A commutative ring* R *is called* ***Noetherian*** *if every ideal of* R *is finitely generated.*

EXAMPLE 6.14. *Thus, by definition, Principal Ideal Domains, like* $\mathbb{Z}, F[x]$, *where* F *is a field, are Noetherian.*

COROLLARY 6.8. *Let R be a commutative ring which is furthermore* Noetherian. *Then every ideal I of R is contained in a maximal ideal.*

PROOF. If I is not maximal itself, then it is properly contained in a proper ideal I_2, and if I_2 is not maximal, it is properly contained in a proper ideal I_3, and so on. Thus, by the ascending chain condition, the chain $I = I_1 \subset I_2 \subset \cdots$ must be finite; therefore I_k must be maximal for some k. □

REMARK 6.13. *The point of this corollary is that we never need the axiom of choice, as long as we work in a Noetherian ring. In fact, Krull (1899 - 1971) proved that*

THEOREM 6.7. *(Krull) Every non-unit ideal in a commutative ring is contained in a maximal ideal.* □

But as far as we know, its proof requires the application of Zorn's lemma, which is equivalent to the axiom of choice. We can also prove Krull's theorem quite easily, but we decide not to provide its proof here.

Now, we are ready to prove:

PROPOSITION 6.5. *Let M be a finitely generated module over a* Noetherian *ring. Then every submodule of M is finitely generated.*

REMARK 6.14. *We note that we have used this proposition in the proof of Theorem 6.6.*

PROOF. First, suppose that we have proved the proposition in the case of $M = R^m$ for any m. Then for any finitely generated R-module M, we have a canonical surjective homomorphism of R-modules: φ: $R^m \longrightarrow M$ (by Theorem 6.1). Given a submodule N of M, one can easily check that its preimage, $L = \varphi^{-1}(N)$, is a submodule of R^m. Then, L is finitely generated by our assumption. Then since the composition map: $L \longrightarrow N$ is automatically surjective, N is finitely generated by Lemma 6.5.

Thus, we are left to prove the proposition when $M = R^m$. For the proof, we use induction on m.

When $m = 1$, the proposition is true because R is Noetherian and submodules of R and ideals of R are the same.

Now suppose $m > 1$. We then consider the projection

$$\pi\colon R^m \longrightarrow R^{m-1}$$

defined by dropping the last entry: $\pi(a_1, ..., a_m) = (a_1, ..., a_{m-1})$. Then its kernel is $\{(0, ..., 0, a_m)\} \approx R$. Now let N be a submodule of R^m, and let φ: $N \longrightarrow R^{m-1}$ be the restriction of π to N. Then by induction hypothesis its image $\varphi(N)$ is a finitely generated submodule of R^{m-1}. Moreover,

$$\ker(\varphi) = \ker(\pi) \cap N$$

is a submodule of $\ker(\pi) \approx R$; hence it is finitely generated too. Therefore, by Lemma 6.5 N is finitely generated, as required. □

As a corollary of Lemma 6.5, we have:

PROPOSITION 6.6. *Let R be a Noetherian ring, and I be an ideal of R. Then the quotient ring R/I is also Noetherian.*

PROOF. Let $\bar{J}$ be an ideal of R/I, and let $J = \pi^{-1}(\bar{J})$ be the corresponding ideal of R (containing I), where

$$\pi\colon R \longrightarrow R/I$$

is the canonical map. Then by hypothesis, J is finitely generated, say by $(a_1, ..., a_m)$. Then by Lemma 6.5, the finite set $(\bar{a}_1, ..., \bar{a}_m)$ generates $\bar{J}$. □

Having said that, we are going to prove Hilbert Basis Theorem: *Every polynomial ring $R[x]$ over a Noetherian commutative ring is Noetherian,* which is one of his famous theorems.

Let us warm up the situation when R is a field first. The following actually proves that the polynomial ring $F[x]$ over a field F is in fact a Principal Ideal Domain (*PID*):

COROLLARY 6.9. *The polynomial ring $F[x]$ over a field F is Noetherian.*

PROOF. Let I be an ideal in $F[x]$. Since we have the degree function $d\colon F[x] \longrightarrow \mathbb{N} \cup \{0\}$, this allows us to choose a non-zero polynomial $f(x) \in I$ of lowest degree among all the non-zero polynomials in I, say

$$f(x) = a_n x^n + \cdots + a_1 x + a_0, \qquad a_n \neq 0,$$

and let

$$g(x) = b_m x^m + \cdots + b_1 x + b_0, \qquad b_m \neq 0$$

be any non-zero polynomial in I. Then the degree m of $g(x)$ is at least n by our choice of $f(x)$. By induction hypothesis, we assume that all the polynomials in I of degree $< m$ are divisible by $f(x)$. We then note that

$$g(x) - (b_m/a_n)x^{m-n} f(x) = g_1(x) \tag{6.13}$$

is an element of I of degree $< m$, and hence it is divisible by induction hypothesis. Therefore, $g(x)$ is divisible by $f(x)$. □

In the above proof, we note that (6.13) is the first step in the division of $g(x)$ by $f(x)$ with remainder. But, we cannot extent this method directly to an arbitrary ring R because division with remainder requires that the leading coefficient of f be a unit, i.e., an invertible element, of R, or we need to know in advance that a_n divides b_m in the ring R, and this is in general not the case. In order to overcome this difficulty, it turns out that we generally need more generators for I than just one generator.

Here is one of Hilbert's famous theorems:

THEOREM 6.8. *(Hilbert Basis Theorem) If a commutative ring R is Noetherian, then so is the polynomial ring $R[x]$.*

In what follows we are going to present 3 different proofs of Hilbert Basis Theorem, of which the first and second "direct" proofs are quite similar, but the third "indirect" proof by a method of contradiction is probably the shortest proof for his theorem.

For the first proof, we need two Lemmas.

LEMMA 6.6. *Let R be a commutative ring. Then the set L of leading coefficients of all the polynomials in an ideal I of a polynomial ring $R[x]$, together with* 0, *forms an ideal of R.*

PROOF. If $\alpha \in L$, then it is equal to the leading coefficient a_n of some polynomial $f(x)$ in I; hence for any $r \in R$, $r\alpha$ is clearly the leading coefficient of the polynomial $rf(x)$, which is automatically in I, unless $r\alpha = 0$ by chance. In both cases, $r\alpha \in L$. Now, if $\beta \in L$, then it is equal to the leading coefficient b_m of some polynomial $g(x) \in I$ of degree, say, $m \geq n = \deg(f(x))$. Then since α is also the leading coefficient of $x^{m-n}f(x) \in I$, $\alpha + \beta$ is the leading coefficient of the polynomial

$$h(x) := x^{m-n}f(x) + g(x) \in I,$$

unless $\alpha + \beta = 0$. In both cases, $\alpha + \beta \in L$. This proves that L is an ideal in R. □

LEMMA 6.7. *Let R be a commutative ring, let P_n denote the set of polynomials $f(x)$ in $R[x]$ with* $\deg(f(x)) < n$, *together with* 0, *and let $S_n = I \cap P_n$, where I is a polynomial ideal of $R[x]$. Then S_n is an R-submodule of the R-module P_n.*

EXERCISE 6.15. *Prove that this lemma is true.*

THE FIRST PROOF OF THEOREM 6.8. Assume that R is a Noetherian ring, and let I be an ideal of $R[x]$. By Proposition 6.4 it is enough to prove that I is finitely generated. By Lemma 6.6, the set L of leading coefficients of the polynomials in I, together with 0, is an ideal of the Noetherian ring R, so there exists a finite set of generators, say, $(\alpha_1, ..., \alpha_k)$, for I. Then we can choose for each $i, 1 \leq i \leq k$, a polynomial $f_i(x) \in I$ with leading coefficient α_i, and we multiply these polynomials by some powers of x if necessary, in order for all of them to have the same common degree, say, n. We also note that the same α_i is the leading coefficient of the new $f_i(x)$ of the common degree n. (From now on until we finish this proof, this n is fixed.)

We then note: since this n is not in general going to be the smallest degree of polynomials in I, the ideal generated by these f_i's is not in general going to be the ideal I.

Nonetheless, Lemma 6.7 says something: We recall that the R-submodule P_n of $R[x]$ is finitely generated by the monomials

$$1, x, x^2, ..., x^{n-1}.$$

Then since R is Noetherian, we may apply Lemma 6.7 and Proposition 6.5 to conclude that its R-submodule S_n is again finitely generated. Say the R-module S_n has a finite set of generators $\{h_1(x), ..., h_s(x)\}$ with each $h_j \in P_n$; hence each $\deg(h_j(x)) < n$ always. We then have

Claim. $(f_1(x), ..., f_k(x); h_1(x), ..., h_s(x)) = I$, that is, the polynomial ideal I is finitely generated by these polynomials, $f_i(x)$'s and $h_j(x)$'s.

PROOF OF CLAIM. Only for simplicity, we denote the ideal

$$(f_1(x), ..., f_k(x); h_1(x), ..., h_s(x))$$

by J. The by our construction, $J \subseteq I$, and hence we only need to prove the opposite inclusion.

For this, let any $g(x) \in I$ be of degree m, and we assume, as an induction hypothesis on m, that all the element of I of degree $< m$ are in the ideal J. If $m < n$ then $g(x) \in I \cap P_n = S_n$; hence $g(x)$ is a linear combination of the $h_j(x)$'s with coefficients in R; consequently $g(x) \in J$. Otherwise, let b be the leading coefficients b_m of the polynomial $g(x)$. Then $b \in L$, the set of all leading coefficients of polynomials in I, so that it is a linear combination of the α_i's, say $b = r_1\alpha_1 + \cdots + r_k\alpha_k$ for some $r_i \in R$. Then since each α_i is the leading coefficient of $f_i(x)$, we see that the polynomial

$$p(x) := x^{m-n} \cdot \sum_{i=1}^{k} r_i f_i(x)$$

has the *same* leading coefficient b and the *same* degree m as $g(x)$; and *moreover* it belongs to the ideal J. Thus,

$$g_1(x) := g(x) - p(x)$$

has degree less than m; hence by induction hypothesis, it is an element of J. Therefore, $g(x) \in J$. □

This finishes the first proof of Hilbert Basis Theorem. □

Here is the second proof of Hilbert Basis Theorem:

THE SECOND PROOF OF THEOREM 6.8. Let R be a Noetherian ring and let I be an ideal of $R[x]$. Then by Proposition 6.4 it is enough to prove that I is finitely generated. For each $n \geq 0$, we define $J_n = \{r \in R : r$ is the leading coefficient of a polynomial in I of degree $n\} \cup \{0\}$. Then it is easy to see that J_n is an ideal of R and that

$$J_n \subset J_{n+1},$$

for each $n \geq 0$, for, for any polynomial $f(x)$, $x \cdot f(x)$ has the same leading coefficient as $f(x)$ does. Since R is Noetherian, there exists N such that $J_N = J_n$ for $n \geq N$. Again by Proposition 6.4, each ideal J_i of R is finitely generated, say,

$$J_i = (r_{i1}, ..., r_{it_i}), \qquad r_{ij} \in R.$$

Now for $i = 1, ..., N$, and $j = 1, ..., t_i$, we can choose, by our construction, a polynomial $f_{ij}(x)$ in I of degree i with leading coefficient r_{ij}. Then we clearly see that the ideal

$$(f_{ij}(x) : 1 \leq i \leq N, 1 \leq j \leq t_i) \subset I,$$

and thus we need to prove the opposite inclusion.

Only for simplicity, we denote the ideal

$$(f_{ij} : 1 \leq i \leq N, 1 \leq j \leq t_i)$$

by I^*. Let any $g(x) \in I$ be of degree n. If $g(x) \equiv 0$, or $n = 0$, then $g(x) \in J_0$; hence it is in I^*. Now for $n > 0$, we assume, as an induction hypothesis on n, that all the elements of I of degree $< n$ are elements of I^*. Let r be the leading coefficient of $g(x)$.

CASE 1. If $n \leq N$, then, since $r \in J_n$, we have

$$r = \sum_{j=1}^{t_n} s_j r_{nj},$$

for some $s_j \in R$. Then the polynomial

$$p(x) := \sum_{j=1}^{t_n} s_j f_{nj}(x)$$

is of degree n, has leading coefficient r, and is an element of I^*. Thus, the polynomial $g(x) - p(x)$ in I has degree $< n$. Thus, by induction hypothesis on n, $g(x) - p(x) \in I^*$; consequently $g(x) \in I^*$. □

CASE 2. Now if $n > N$, then $r \in I_n = I_N$; hence

$$r = \sum_{j=1}^{t_N} s_j r_{Nj},$$

for some $s_j \in R$. Then the polynomial

$$p(x) := \sum_{j=1}^{t_N} s_j x^{n-N} f_{Nj}(x)$$

is of degree n, has leading coefficient r, and is an element of I^*. Thus, the polynomial $g(x) - p(x)$ in I has degree $< n$. Thus, by induction hypothesis on n, $g(x) - p(x) \in I^*$; consequently $g(x) \in I^*$. □

This finishes the second proof of Hilbert Basis Theorem. □

Now, here is the third proof of Hilbert Basis Theorem:

THE THIRD PROOF OF THEOREM 6.8. In what follows, we are going to prove:

If $R[x]$ is not Noetherian, then R is not a Noetherian ring.

Let I be an ideal of $R[x]$ that is not finitely generated. Let $f_1 \in I$ be a polynomial of least degree. If f_k, $k \geq 1$, has already been chosen, let f_{k+1}

be a polynomial of least degree in $I - (f_1, ..., f_k)$. For these polynomials f_k, let n_k be the degree of f_k and $a_k \in R$ be the leading coefficient of f_k, for $k = 1, 2, ...$ Then by our choice of f_k, we see that

$$n_1 \leq n_2 \leq \cdots.$$

Moreover, the increasing chain of ideals in R,

$$(a_1) \subset (a_1, a_2) \subset \cdots$$

must be a strictly increasing chain of ideals in R. For, if $(a_1, ..., a_k) = (a_1, ..., a_{k+1})$, then

$$a_{k+1} = \sum_{i=1}^{k} b_i a_i, \qquad b_i \in R$$

and hence the polynomial

$$p := f_{k+1} - \sum_{i=1}^{k} b_i x^{n_{k+1} - n_i} f_i \in I - (f_1, ..., f_k)$$

is of lower degree than f_{k+1}, contradicting our choice of f_{k+1}, and this finishes the third proof of Hilbert Basis Theorem. □

The Hilbert Basis Theorem shows by induction that

COROLLARY 6.10. *The polynomial ring $R[x_1, ..., x_n]$ in several variables over a Noetherian ring R is Noetherian.* □

Hence the rings $\mathbb{Z}[x_1, ..., x_n]$ and $F[x_1, ..., x_n]$ where F is a field are Noetherian. Also, quotients of Noetherian rings are Noetherian.

Combining Proposition 6.6 with the Hilbert Basis Theorem, we obtain:

COROLLARY 6.11. *Any commutative ring which is a quotient of a polynomial ring over the integers or over a field is Noetherian.* □

Exercises for Section 6.4

EXERCISE 6.16. *Let S be a subset of $\mathbb{C}^n$.*

(1) *Prove that the set $I(S)$ of those polynomials $f(x_1, ..., x_n)$ vanishing on S is an ideal of $\mathbb{C}[x_1, ..., x_n]$.*

(2) *Prove that there is a finite set of polynomials*

$$(f_1(x_1,...,x_n),...,f_k(x_1,...,x_n))$$

such that any polynomial vanishing identically on S is a linear combination of this set, with polynomial coefficients.

EXERCISE 6.17. *Let $V \subset \mathbb{C}^n$ be the locus of zeros of an infinite set of polynomials in n variables:*

$$\begin{cases} f_1(x_1,...,x_n) &= 0, \\ f_2(x_1,...,x_n) &= 0, \\ \quad\cdots & \\ \quad\cdots & \end{cases}$$

Prove that there exist a finite subset of these polynomials whose zeros define the same locus.

5. The Structure Theorem for Abelian Groups

In this section we apply what we have developed in the previous sections to the structure theory of abelian groups: *A finitely generated abelian group M is a direct sum of cyclic groups.* To establish it, we are going to apply Theorem 6.3, and others which are its variants and Proposition 6.3. More precisely, since any abelian group is a $\mathbb{Z}$-module, we know by Theorem 6.3 that there exists a diagonal presentation matrix for M, and a proper application of Proposition 6.3 will enable us to interpret the meaning of this diagonal matrix for an abelian group. This means in part that we need to introduce a few more terminologies for modules in order to have a better understanding of the subject.

DEFINITION 6.17. *Let $W_1,...,W_k$ be submodules of an R-module M.*

(1) *Their* ***sum*** *is the submodule which they generate, i.e.,*

$$W_1 + \cdots + W_k := \{w_1 + \cdots + w_k : \text{ each } w_i \in W_i\}.$$

(2) *We say that M is the* ***direct sum,***

$$M = W_1 \bigoplus \cdots \bigoplus W_k,$$

of submodules W_i of M if

(a) *they generate:* $M = W_1 + \cdots + W_k$;

(b) *they are independent: If* $w_1 + \cdots + w_k = 0$, *with each* $w_i \in W_i$, *then each* $w_i = 0$ *for* $i = 1, ..., k$.

(3) *The* ***direct product*** *of submodules W_i is the product set*

$$W_1 \times \cdots \times W_k = \{(w_1, ..., w_k) : \ w_i \in W_i\}$$

together with vector addition and scalar multiplication:

$$\begin{aligned} (w_1, ..., w_k) + (w_1', ..., w_k') &= (w_1 + w_1', ..., w_k + w_k'), \\ r(w_1, ..., w_k) &= (rw_1, ..., rw_k). \end{aligned}$$

Then it is very easy to check that the direct product $W_1 \times \cdots \times W_k$ is an R-module under these operations. It is also very easy to see that

LEMMA 6.8. *The sum of submodules W_i is a submodule of M and it is the smallest submodule of M among those submodules containing each W_i.* □

It is also very easy to check that

LEMMA 6.9. *M is the direct sum of the submodules W_i if every element $v \in M$ can be written uniquely in the form*

$$v = w_1 + \cdots + w_k,$$

with each $w_i \in W_i$. □

LEMMA 6.10. *Two submodules W_1 and W_2 are independent if and only if $W_1 \cap W_2 = \{0\}$.* □

Finite direct products and finite direct sums are isomorphic, as the following proposition shows:

PROPOSITION 6.7. *Let $W_1, ..., W_k$ be submodules of an R- module M. Then*

a: *The map* σ: $W_1 \times \cdots \times W_k \longrightarrow M$ *defined by sending each* $(w_1, ..., w_k) \mapsto w_1 + \cdots + w_k$ *is a homomorphism of R-modules, and its image is the sum* $W_1 + \cdots + W_k$.

b: *The homomorphism σ is an isomorphism if and only if M is the direct sum of the submodules W_i.*

EXERCISE 6.18. *The proof of this proposition is very easy, so we leave its proof as an exercise.*

THEOREM 6.9. *Let M be a finitely generated abelian group. Then M is a direct sum of finite cyclic subgroups $C_{d_1}, ..., C_{d_k}$ and a free abelian group L of finite rank:*

$$M = C_{d_1} \bigoplus \cdots \bigoplus C_{d_k} \bigoplus L,$$

where the order, i.e., the cardinality, d_i of each C_{d_i} is greater than 1, and $d_1|d_2|d_3....$

REMARK 6.15. *Since we are dealing with subgroups of an abelian group, we will use as usual additive notation for the law of composition in the cyclic group. Namely, C_n denotes a cyclic group generated by a single element v with one relation $nv = 0$. Therefore,*

$$C_n \text{ is obviously isomorphic to } \mathbb{Z}/n\mathbb{Z},$$

where by definition sending the residue of an integer r to rv is the isomorphism: $\mathbb{Z}/n\mathbb{Z} \longrightarrow C_n$.

PROOF. Since $\mathbb{Z}$ is a Noetherian ring and M is a finitely generated $\mathbb{Z}$-module, we can apply the theory of generators and linear relations among them to have a presentation matrix A for M, which is determined by a set of generators and a complete set of relations. Thus, by Proposition 6.3, the matrix A may be replaced by QAP^{-1}, where Q and P are invertible integer matrices. Thus, we may assume that A is diagonal, that the diagonal entries are non-zero, and that each diagonal entry divides the next one. Moreover, thanks to Proposition 6.3 we can drop any column of zeros, and any row and column in which the diagonal entry, i.e., the pivot, is 1. Thus, we may further assume that the diagonal entries d_i of the matrix A are neither 0 nor 1:

$$A = \begin{bmatrix} d_1 & & & & \\ & d_2 & & & \\ & & \cdot & & \\ & & & \cdot & \\ & & & & d_k \\ & & 0 & & \end{bmatrix} \tag{6.14}$$

Thus, we may assume that the size of A is $m \times k$ for some k and m with $k \leq m$. In terms of generators and linear relations, the matrix A means that M is generated by m elements $v_1, ..., v_m$ satisfying a complete set of k relations among these generators:

$$d_1 v_1 = 0, \ d_2 v_2 = 0, \ ..., d_k v_k = 0.$$

Thus, the rest of the generators, $v_{k+1}, ..., v_m$ are free, meaning that we cannot do anything on them because there does not exist any linear relations among them. This is the reason why we need to figure out the complete set of relations among the generators.

For $j = 1, ..., k$, we denote by C_j the cyclic subgroup generated by v_j satisfying $d_j v_j = 0$. We also denote by L the subgroup of M generated by the remaining free generators $v_{k+1}, ..., v_m$. Since they never appear in the set of complete k relations, there is no linear relations among them, and hence L is a free abelian group of rank $m - k$.

To finish the proof, we need to verify that

$$M = C_1 \bigoplus \cdots \bigoplus C_k \bigoplus L,$$

and that each C_j is a cyclic group of order d_j.

First, it is very clear that M is a sum of these subgroups since all the generators v_i's appear either one of them, C_j or L. To show that they are independent, we suppose that we have a relation:

$$z_1 + \cdots + z_k + w = 0,$$

where each $z_j \in C_j$ and $w \in L$. Since each C_j is a cyclic subgroup with generator v_j and since L is a free abelian subgroup of M, we can write each $z_j = r_j v_j$ for some integer $r_j \in \mathbb{Z}$ and each $w = r_{k+1}v_{k+1} + \cdots + r_m v_m$ for some integers $r_i \in \mathbb{Z}$. We thus have a linear relation among the generators $v_1, ..., v_m$:

$$r_1 v_1 + \cdots + r_m v_m = 0.$$

Then since we already know a complete set of relations, the integer column vector $(r_1, ..., r_m)^t$ is a linear combination of the columns of the matrix A; hence for $j \geq k+1$, we see that $r_j = 0$. Thus, $w = 0$. In addition to this, for $j \leq k$, each r_j must be divisible by d_j, say, $r_j = s_j d_j$ for some integer $s_j \in \mathbb{Z}$. Thus, each $z_j = s_j d_j v_j = 0$, and this proves that both the subgroups are independent and the order of each cyclic subgroup C_j is d_j. □

COROLLARY 6.12. *A finite abelian group M is a direct sum of finite cyclic subgroups $C_{d_1}, ..., C_{d_k}$.*

PROOF. Since it is finitely generated and since it is a finite set, we cannot possibly have a subgroup consisting of infinitely many elements, so that by Theorem 6.9 we have

$$M = C_{d_1} \bigoplus \cdots \bigoplus C_{d_k},$$

where $1 < d_1 | d_2 | d_3 ....$ □

It is sometimes convenient to decompose the cyclic groups further into cyclic groups of prime power order. For this we need:

LEMMA 6.11. *Let r, s be two relatively prime integers. Then the cyclic group C_{rs} of order rs is isomorphic to the direct sum of cyclic groups of order r and order s, respectively,*

$$C_{rs} \xrightarrow{\sim} C_r \bigoplus C_s.$$

PROOF. By Proposition 6.7 we already know that $C_r \bigoplus C_s$ is isomorphic to the direct product $C_r \times C_s$. Thus, we are going to establish an isomorphism

$$C_{rs} \xrightarrow{\sim} C_r \times C_s.$$

For this, let v be a generator for C_{rs} and w, z be generators of C_s and C_s, respectively. Then the map sending $kv \mapsto (kw, kz)$ for $0 \leq k < rs$, is clearly a Z-module homomorphism with kernel 0. Then since both C_{rs} and $C_r \times C_s$ have the same cardinality rs, they are bijective. □

COROLLARY 6.13. *Moreover, if these cyclic groups C_{rs}, C_r, and C_s, where $(r, s) = 1$, are subgroups of an abelian group M, then $C_{rs} = C_r \bigoplus C_s$.*

PROOF. Since both C_r and C_s are clearly subgroups of C_{rs}, the smallest subgroup $C_r \bigoplus C_s$ containing both C_r and C_s is a subgroup of C_{rs}. But then since they have the same cardinality, they must be the same. □

Combining this result with the Structure Theorem, Theorem 6.9, we get

COROLLARY 6.14. *(Structure Theorem, alternate form) Every finitely generated abelian group is a direct sum of cyclic groups of prime power orders and of a free abelian group of finite rank.* □

THEOREM 6.10. *(Uniqueness for the Structure Theorem)*

- **a:** *Suppose that a finite abelian group M is a direct sum of cyclic groups $C_{d_1} \bigoplus \cdots \bigoplus C_{d_k}$, where $1 < d_1|d_2|d_3\ldots.$ Then the integers d_i are (uniquely) determined by the group M.*
- **b:** *The same is true if the decomposition is into prime power orders, that is, if each d_j is the power of a prime.*

EXERCISE 6.19. *We leave the proof of this theorem as an exercise.*

EXAMPLE 6.15. *If the order of a finite abelian group is square free, i.e., a product of distinct primes, then there is no problem since we have just one decomposition. For instance if the order is*

$$30 = 2 \times 3 \times 5,$$

then M must be isomorphic to $C_2 \bigoplus C_3 \bigoplus C_5$. Otherwise, we can have several non-isomorphic decompositions. For instance, if the order is

$$16 = 2^4,$$

then $C_2 \bigoplus C_2 \bigoplus C_4 \xrightarrow{\sim} C_2 \times C_2 \times C_4$ is never isomorphic to $C_4 \bigoplus C_4 \xrightarrow{\sim} C_4 \times C_4$, because the Klein 4 group $C_2 \times C_2$ is never isomorphic to the cyclic group C_4, for instance, and hence the two decompositions are not isomorphic.

REMARK 6.16. *This examples suggests that the counting of elements is sometimes easier if we use Proposition 6.7.*

Now, since a cyclic group C_d of order d is isomorphic to the standard cyclic group $\mathbb{Z}/d\mathbb{Z}$, Proposition 6.7 allows us to express the Structure Theorem as follows:

THEOREM 6.11. *(Product version of the Structure Theorem) Every finitely generated abelian group M is isomorphic to a direct product of cyclic groups*

$$\mathbb{Z}/d_1\mathbb{Z} \times \cdots \times \mathbb{Z}/d_k\mathbb{Z} \times \mathbb{Z}^r,$$

where d_i, r are integers. There is a decomposition in which each d_i divides the next one, and there is another decomposition in which each d_i is a prime power. □

We note that the most closest rings to $\mathbb{Z}$ are Euclidean domains in the sense that both of them are a *PID* (Principal Ideal Domain) and hence Noetherian. For instance,

EXAMPLE 6.16. *The polynomial ring $F[t]$ in one variable t over a field F is a Euclidean domain.*

The classification of abelian groups can carry over to Euclidean domains without essential change: Since Euclidean domain R is Noetherian, any finitely generated R-module M has a presentation matrix in (6.12), and then by the diagonalization theorem (Theorem 6.4) there is a presentation matrix A which is diagonal.

In order to carry along the analogy with abelian groups, we definitely need:

DEFINITION 6.18. *A* ***cyclic*** $\boldsymbol{R}$***-module*** M *is an* R*-module which is generated by a single element* v.

COROLLARY 6.15. *Let* M *be an* R*-module. Then* M *is cyclic generated by* v *if and only if it is isomorphic to a quotient module* R/I, *where* I *is the ideal of* R *consisting of those elements* α *such that* $\alpha v = 0$.

PROOF. By Theorem 6.1, there exists a surjective homomorphism of R-modules $\varphi\colon R \longrightarrow M$ which sends $r1 \mapsto rv$. Then the ideal I is nothing but the kernel of φ; hence M is isomorphic to R/I. Conversely, if an R-module M is isomorphic to R/I, then the image of the generator 1 for the quotient ring R/I under such an isomorphism will generate M. □

Thus, if R is a Euclidean domain, then the ideal I in this corollary is a principal ideal (α), so that any cyclic module M over a Euclidean domain R is isomorphic to $R/(\alpha)$ for some $\alpha \in R$.

We will need the following theorem in the next section:

THEOREM 6.12.

a: Let M be a finitely generated module over a Euclidean domain R. Then M is a direct sum of cyclic R-modules C_j and a free R-module L. More precisely, there is an isomorphism

$$\varphi\colon M \xrightarrow{\sim} R/(d_1) \times \cdots \times R/(d_k) \times R^r,$$

of M with a direct product of cyclic R-modules $R/(d_i)$ and a free R-module R^r, where r is non-negative, the elements $d_1, ..., d_k$ are not units and not zeros, and $d_i | d_{i+1}$ for each $i = 1, ..., k-1$.

b: The same assertion as part a, except that the condition that d_i divides d_{i+1} is replaced by the following one: Each d_i is a power of a prime element of R. Thus M is isomorphic to a product of the form

$$R/(p_1^{e_1}) \times \cdots \times R/(p_k^{e_k}) \times R^r,$$

where repetition of primes are allowed. □

EXAMPLE 6.17. *Let M be the $F[t]$-module presented by the matrix A of Example 6.7.*

$$A = \begin{bmatrix} (t-2)(t-1) & t-2 \\ (t-1)^3 & (t-2)(t-1) \end{bmatrix}.$$

Then part (1) of Proposition 6.3 tells us that M is also presented by the diagonal matrix

$$A' = \begin{bmatrix} 1 & 0 \\ 0 & (t-2)(t-1)^2 \end{bmatrix}.$$

Again by other parts of Proposition 6.3 we can drop the first row and column from A' because the $(1,1)$ entry of A' has the pivot 1. Thus, M is represented by the 1×1 matrix $[g(t)]$, where

$$g(t) = (t-1)^2(t-2).$$

Therefore, M is a cyclic module, which is isomorphic to $F[t]/(g(t))$. Furthermore, since $g(t)$ is the product of two relatively prime factors, we can further decompose it; consequently it is isomorphic to the direct product of two cyclic modules:

$$M \approx F[t]/(g(x)) \approx F[t]/(t-1)^2 \times F[t]/(t-2).$$

REMARK 6.17. *We note that Principal Ideal Domains (PID's) are in some sense the next closest ring to $\mathbb{Z}$, and hence with slightly more work, we can extend Theorem 6.12 to modules over any PID.*

REMARK 6.18. *Over a PID, it is also true that the prime powers occurring in part b of Theorem 6.12 are unique up to unit factors. A substitute for the counting argument including Proposition 6.7, which proves Theorem 6.10, must be found to prove this fact.*

Exercises for Section 6.5

EXERCISE 6.20. *Determine the number of non-isomorphic abelian groups of order $400 = 2^4 \cdot 5^2$.*

EXERCISE 6.21. *Find an isomorphic direct product of cyclic groups, when M is the abelian group generated by x, y, and z, with the following relations:*

(1) $3x + 2y + 8z = 0, 2x + 4z = 0.$
(2) $x + y = 0, 2x = 0, 4x + 2z = 0, 4x + 2y + 2z = 0.$
(3) $2x + y = 0, x - y = 3z = 0.$
(4) $2x - 4y = 0, 2x + 2y + z = 0.$

EXERCISE 6.22. *Find a direct sum of cyclic groups which is isomorphic to the abelian group presented by the matrix*

$$A = \begin{bmatrix} 3 & 3 & 3 \\ 3 & 3 & 0 \\ 3 & 0 & 3 \end{bmatrix}.$$

EXERCISE 6.23. *Let R be a ring, and let M be an R-module, presented by a diagonal $m \times n$ matrix A:*

$$R^m / AR^n \approx M.$$

Let $(v_1, \ldots, v_m)$ be the corresponding generators of M, and let d_i be the diagonal entries of A. Prove that M is isomorphic to a direct product of the modules $R/(d_i)$.

6. $F[t]$-Modules and Endomorphisms on a Vector Space

We have so far in this chapter extended the method developed first for abelian groups to modules over Euclidean domains. As we shall see soon, it turns out that a theory on an $F[t]$-module can be equivalently described as a theory on a vector space with an endomorphism, and conversely. But, we first develop the theory without talking about $F[t]$-modules and this will provide the reader with a chance to see how polynomials in one variable can play its role in our study of endomorphisms on a vector space.

6.1. The Minimal Polynomial of an Endomorphism. We recall that similar matrices, PAP^{-1} with invertible matrices P, have the same determinant and the same trace as the square matrix A. These invariants take values in the field of scalars and hence they are sometimes called "**numerical invariants**". A different kind of invariant for similarity is the function $F^{n^2} \longrightarrow F[\lambda]$ which associates to each $n \times n$ matrix A its characteristic polynomial, $\det(\lambda I - A)$, in one unknown λ over a field F. In this subsection, we are going to construct another such "polynomial invariant", called the "*minimal polynomial*" of A.

DEFINITION 6.19. *We denote by*

$$End(V)$$

the set of all endomorphisms on a vector space V over a field F.

We then easily see that for each scalar α, left multiplication by α is certainly an endomorphism on V, and we denote this endomorphism by

$$\alpha I \colon V \longrightarrow V.$$

Thus, $End(V)$ contains the collection F of scalars as a subset. If $t \colon V \longrightarrow V$ is an endomorphism on V, then its scalar multiple,

$$\alpha t \colon V \longrightarrow V,$$

is again an endomorphism on V, and hence any polynomial expression in t of positive degree is again an endomorphism on V.

In fact, we have

LEMMA 6.12. *Let V be a vector space over a field F.*

- **a:** *$End(V)$ is an F-vector space under pointwise addition and pointwise scalar multiplication.*
- **b:** *$End(V)$ is a ring (usually non-commutative) under the operations of pointwise sum, composition, and selection of the identity map on V as a unit.*
- **c:** *The subring of $End(V)$ generated by an endomorphism*

$$t : V \longrightarrow V$$

and scalar multiples of the identity map I on V is a commutative ring containing F and t.

d: *If V has dimension n, then there exists an isomorphism*

$$End(V) = Hom(F^n, F^n) \approx F^{n\times n}$$

of F-vector spaces.

PROOF. It is very routine to verify them, so we leave their proofs as an exercise. □

In particular, part c of this lemma means that, by the universally repelling property of the polynomial ring, there exists a unique ring homomorphism from the polynomial ring:

$$E_t\colon F[x] \longrightarrow End(V)$$

which sends the indeterminate x to the generator t and whose restriction on F is the identity map.

THEOREM 6.13. *Let $t\colon V \longrightarrow V$ be an endomorphism on a* finite *dimensional vector space V over F. Then there exists (unique) monic (not-necessarily irreducible) polynomial $m(x)$ of least degree such that $m(t) = 0$ under*

$$E_t : F[x] \longrightarrow End(V).$$

PROOF. Since $\dim(V) = n$, $End(V) = Hom(F^n, F^n)$ and hence it is isomorphic to the vector space $F^{n\times n}$ of dimension n^2. Thus, for example, the $n^2 + 1$ of endomorphisms on V

$$1, t, ..., t^{n^2}$$

is linearly dependent. This just means that there exists a non-constant polynomial $f(x) \in F[x]$ such that $f(t) = 0$. That is, the kernel of the evaluation homomorphism E_t is bigger than (0), hence since $F[x]$ is a *PID*, the kernel is generated by a single (not-necessarily irreducible) polynomial $m(x)$ over F. If we require this polynomial $m(x)$ to be monic, then it is uniquely determined. □

DEFINITION 6.20. *This not-necessarily irreducible polynomial $m(x)$ over F is called the* **minimal polynomial** *of $t\colon V \longrightarrow V$.*

COROLLARY 6.16. *Let V be a finite-dimensional vector space over a field F. Then the minimal polynomial of an endomorphism $t \in End(V)$ is identical to the minimal polynomial of any matrix representing t.*

PROOF. Since the subring of the matrix ring $F^{n\times n}$ generated by left multiplication by an $n \times n$ matrix A: $F^n \longrightarrow F^n$ and scalar multiples of the identity map I on F^n is a commutative ring containing F and A, we have exactly the same theorem for the ring $F^{n\times n}$ of square matrices as Theorem 6.13, and hence we have the minimal polynomial of A, and then since for each basis $\mathbf{B}$ of V the function:

$$End(V) \longrightarrow F^{n\times n}$$

sending each endomorphism $t \in End(V)$ to the matrix A of t with respect to $\mathbf{B}$ is an isomorphism of rings, we have the desired result. □

COROLLARY 6.17. *Similar matrices have the same minimal polynomial.*

PROOF. This is because similar matrices represent the same endomorphism relative to different bases. □

6.1.1. *Nilpotent Endomorphisms.*

DEFINITION 6.21. *An endomorphism t: $V \longrightarrow V$ is called **nilpotent** if there exists positive integer k such that $t^k = 0$.*

Thus, the minimal polynomial of a nilpotent endomorphism t is

$$x^n$$

for some n.

In particular if $\dim(V) = n =$ the degree of the minimal polynomial of a nilpotent endomorphism t on V, then there exists a vector $w \in V$ such that $t^{n-1}(w) \neq 0$; then the n vectors of V,

$$w, t(w), ..., t^{n-1}(w),$$

are linearly independent. For, if $0 = c_0w + c_1t(w) + \cdots + c_{n-1}t^{n-1}(w)$, say, with the first non-zero coefficient c_j, then $0 = t^{n-j-1}(c_jt^j(w)) = c_jt^{n-1}(w)$, which is a contradiction to $t^{n-1}(w) \neq 0$. Thus, if we take them as a basis for V, we have

$$t(w, t(w), ..., t^{n-1}(w)) = (w, t(w), ..., t^{n-1}(w))A,$$

where

$$A = \begin{bmatrix} 0 & 0 & .. & 0 & 0 \\ 1 & 0 & .. & 0 & 0 \\ 0 & 1 & .. & 0 & 0 \\ . & . & & . & . \\ & & .. & & \\ . & . & & . & . \\ 0 & 0 & .. & 0 & 0 \\ 0 & 0 & .. & 1 & 0 \end{bmatrix}.$$

This $n \times n$ matrix surely has the minimal polynomial x^n; and moreover we can easily see that its characteristic polynomial is also x^n, which is a special case of the fact that the minimal polynomial of t divides the characteristic polynomial of t.

6.1.2. ***Companion Matrix.*** Now suppose we are given a monic polynomial of degree m,

$$g(x) = g_0 + g_1 x + \cdots g_{m-1} x^{m-1} + x^m.$$

From the above arguments for nilpotent endomorphisms, we can easily construct an $m \times m$ matrix:

$$A_g = \begin{bmatrix} 0 & 0 & .. & 0 & -g_0 \\ 1 & 0 & .. & 0 & -g_1 \\ 0 & 1 & .. & 0 & . \\ . & . & & . & . \\ & & .. & & \\ . & . & & . & . \\ 0 & 0 & .. & 0 & -g_{m-2} \\ 0 & 0 & .. & 1 & -g_{m-1} \end{bmatrix}.$$

DEFINITION 6.22. *A_g is called the* ***companion matrix*** *of the monic polynomial,*

$$g(x) = g_0 + g_1 x + \cdots g_{m-1} x^{m-1} + x^m$$

in $F[x]$.

PROPOSITION 6.8. *The companion matrix* A_g *of a monic polynomial*

$$g(x) = g_0 + g_1 x + \cdots g_{m-1} x^{m-1} + x^m$$

of degree m *is an* $m \times m$ *matrix with minimal polynomial* $g(x)$ *and characteristic polynomial* $(-1)^m g(x)$.

PROOF. We note that left multiplication by A_g is a linear transformation on the spaceF^m, $t : F^m \longrightarrow F^m$, and it sends the standard basis elements ε_j of F^m as follows:

$$\begin{aligned} t(\varepsilon_j) &= \varepsilon_{j+1} \quad \text{for } 1 \leq j \leq m-1, \\ t(\varepsilon_m) &= -\varepsilon_1 g_0 - \varepsilon_2 g_1 - \cdots - \varepsilon_m g_{m-1}. \end{aligned}$$

Thus, by the first $m-1$ equations, we see that

$$\begin{aligned} \varepsilon_m &= t(\varepsilon_{m-1}) = t^2(\varepsilon_{m-2)} = \ldots = t^{m-1}(\varepsilon_1), \\ \varepsilon_{j+1} &= t^j(\varepsilon_1), \end{aligned}$$

for each $j < m$. Thus, we get

$$t(\varepsilon_m) = t^m(\varepsilon_1),$$

and hence replacing each ε_{j+1} in the second equation with this, we get

$$t^m(\varepsilon_1) = -(g_0 + g_1 t + \cdots + g_{m-1} t^{m-1})(\varepsilon_1).$$

Thus, we see that

$$g(t)(\varepsilon_1) = 0,$$

and consequently we also see that

$$g(t)(\varepsilon_{j+1}) = g(t)(t^j(\varepsilon_1)) = t^j(g(t)(\varepsilon_1)) = 0,$$

for every $j = 1, 2, \ldots, m-1$. Therefore, $g(t)(w) = 0$ for every vector w in F^m; hence we must have

$$g(t) = 0.$$

On the other hand, if $f(x) = \sum_{j=1}^{d} f_j x^j$ is a polynomial of lower degree $d < m$, then, in a similar manner, we get $\left(\sum_{j=1}^{d} f_j t^j\right)(\varepsilon_1) = \ldots = \sum_{j=1}^{d} f_j \varepsilon_{j+1}$, which is a linear combination of the ε_j's; hence it is zero only when every coefficient f_j of $f(x)$ is zero, because the ε_j are the standard basis of F^m. Therefore, $g(x)$ is indeed the minimal polynomial for t and hence for A_g.

Then one can easily check that

$$\begin{aligned}\det(A_g - \lambda I) &= \det \begin{bmatrix} -\lambda & 0 & .. & 0 & -g_0 \\ 1 & -\lambda & .. & 0 & -g_1 \\ 0 & 1 & .. & 0 & . \\ . & . & .. & . & . \\ . & . & & . & . \\ 0 & 0 & .. & -\lambda & -g_{m-2} \\ 0 & 0 & .. & 1 & -\lambda - g_{m-1} \end{bmatrix} \\ &= (-1)^m g(\lambda),\end{aligned}$$

where we get the last equality by the minor expansion of it through the last column. □

EXAMPLE 6.18. *When $m = 4$, we have*

$$\begin{aligned}\det(A_g - \lambda I) &= \det \begin{bmatrix} -\lambda & 0 & 0 & -g_0 \\ 1 & -\lambda & 0 & -g_1 \\ 0 & 1 & -\lambda & -g_2 \\ 0 & 0 & 1 & -\lambda - g_3 \end{bmatrix} \\ &= (-1)^4(g_0 + g_1\lambda + g_2\lambda^2 + g_3\lambda^3 + \lambda^4) \\ &= (-1)^4 g(\lambda).\end{aligned}$$

This matrix is special in the following sense: Let $t \in End(V)$ be a non-trivial endomorphism on a vector space of dimension n. Then we can easily see that there exists a non-zero vector $w \in V$ such that $t(w) \neq 0$, thus we can have a sequence of vectors: $w, t(w), t^2(w), ...$ in V. Let

$$C_w(t)$$

be the subspace of V spanned by these vectors, $w, t(w), t^2(w), \ldots$ We can easily see that

LEMMA 6.13. *For a non-zero vector $w \in V$ of dimension n, $C_w(t)$ is t-invariant, i.e., $t(C_w(t)) \subset C_w(t)$.* □

DEFINITION 6.23. *$C_w(t)$ is called the t-**cyclic subspace** of V.*

Moreover, we have

LEMMA 6.14. *For a non-zero vector $w \in V$ of dimension n, there exist $m \le n = \dim(V)$ and scalars $g_0, g_1, \ldots, g_{m-1} \in F$ such that*

$$\{w, t(w), \ldots, t^{m-1}(w)\}$$

is a basis of the t-cyclic subspace $C_w(t)$ of dimension m, and

$$t(w, t(w), \ldots, t^{m-1}(w)) = (w, t(w), \ldots, t^{m-1}(w))A_g,$$

where

$$g = g_0 + g_1 x + \ldots + g_{m-1}x^{m-1} + x^m$$

is a monic polynomial in x of degree m, and hence

$$A_g = \begin{bmatrix} 0 & 0 & .. & 0 & -g_0 \\ 1 & 0 & .. & 0 & -g_1 \\ 0 & 1 & .. & 0 & . \\ . & . & .. & . & . \\ . & . & & . & . \\ 0 & 0 & .. & 0 & -g_{m-2} \\ 0 & 0 & .. & 1 & -g_{m-1} \end{bmatrix}$$

is the matrix of $t|_{C_w(t)}$ with respect to this basis.

PROOF. We know that $w \neq 0$ itself is linearly independent and since V has dimension n,there exists an integer m such that

$$w, t(w), \ldots, t^{m-1}(w)$$

are linearly independent, but

$$w, t(w), \ldots, t^{m-1}(w), t^m(w)$$

are linearly dependent with $m \leq n$. Thus, we have a linear dependence relation:

$$g_0 w + g_1 t(w) + ... + g_{m-1} t^{m-1}(w) + t^m(w) = 0,$$

for some scalars $g_0, ..., g_{m-1} \in F$. That is,

$$(g_0 + g_1 t + ... + g_{m-1} t^{m-1} + t^m)(w) = 0.$$

Thus, we obtain the matrix A_g of $t_{C_w(t)}$ with respect to this basis of $C_w(t)$, where $g = g_0 + g_1 x + ... + g_{m-1} x^{m-1} + x^m$. □

Moreover, we see that

LEMMA 6.15. *The polynomial $g(x)$ divides another polynomial $f(x)$ if and only if $f(t)(w) = 0$.*

PROOF. If $f(t)(w) = 0$, then we apply the Euclidean algorithm:

$$f(x) = q(x)g(x) + r(x),$$

with $0 \leq \deg(r(x)) < \deg(g(x))$ for unique $q(x)$ and $r(x) = b_0 + b_1 x + ... + b_{m-1} x^{m-1}$. Thus, $f(t)(w) = 0$ implies that $r(t)(w) = 0$, but then since $w, t(w), ..., t^{m-1}(w)$ is linearly independent, each coefficient $b_i = 0$. Thus, $g(x)$ should divide $f(x)$. The converse of the lemma is trivial. □

We thus proved

PROPOSITION 6.9. *Let $t \in End(V)$ be an endomorphism on V of dimension n. Then for any non-zero vector $w \in V$, there exists a monic (not-necessarily irreducible) polynomial $g(x) \in F[x]$ of least degree such that $g(t)(w) = 0$.* □

DEFINITION 6.24. *Such a not-necessarily irreducible polynomial $g(x)$ is called t-**minimum polynomial** of the non-zero vector w.*

LEMMA 6.16. *For the restriction of t on t-invariant subspace $C_w(t)$, $t|_{C_w(t)} \in End(C_w(t))$, g is the characteristic polynomial of $t|_{C_w(t)}$.*

PROOF. Write $f(x)$ be the characteristic polynomial of the matrix A_g of $t|_{C_w(t)}$. Adding x^i multiple of each i-th row to the first row for

$2 \le i \le n$, we obtain

$$
\begin{aligned}
f(x) &= \det \begin{bmatrix} x & 0 & \cdots & 0 & g_0 \\ -1 & x & \cdots & 0 & g_1 \\ 0 & -1 & \cdots & 0 & g_2 \\ \vdots & \vdots & & \vdots & \vdots \\ 0 & 0 & \cdots & x & g_{m-2} \\ 0 & 0 & \cdots & -1 & x+g_{m-1} \end{bmatrix} \\
&= \det \begin{bmatrix} 0 & 0 & \cdots & 0 & g(x) \\ -1 & x & \cdots & 0 & g_1 \\ 0 & -1 & \cdots & 0 & g_2 \\ \vdots & \vdots & & \vdots & \vdots \\ 0 & 0 & \cdots & x & g_{m-2} \\ 0 & 0 & \cdots & -1 & x+g_{m-1} \end{bmatrix}.
\end{aligned}
$$

Thus, the minor expansion through the first row gives:

$$f(x) = (-1)^{m+1} g(x)(-1)^{m-1} = g(x).$$

□

PROPOSITION 6.10. *Let t be an endomorphism on a vector space V of dimension n, W its t-invariant subspace. Then the characteristic polynomial of $t|_W$ divides that of t.*

PROOF. Let $\mathbf{C} = (u_1, ..., u_r)$ be a basis of W and we extend it to a basis of V:

$$\mathbf{B} = (u_1, ..., u_r; v_1, ..., v_s).$$

Then the matrix of t with respect to $\mathbf{B}$ looks like:

$$t((u_1, ..., u_r; v_1, ..., v_s)) = (u_1, ..., u_r; v_1, ..., v_s)M,$$

$$M = \begin{bmatrix} A & B \\ & C \end{bmatrix},$$

where A is the matrix of t_W with respect to $\mathbf{C} = (u_1, ..., u_r)$. Thus, the characteristic polynomial $\det(xI - M)$ of t is expanded as follows:

$$\begin{aligned}\det(xI - M) &= \det\begin{bmatrix} xI_r - A & -B \\ 0 & xI_s - C \end{bmatrix} \\ &= \det(xI_r - A) \cdot \det(xI_s - C).\end{aligned}$$

□

COROLLARY 6.18. *The characteristic polynomial of $t|_{C_w(t)}$, i.e., $g(x)$ constructed in Lemma 6.14, divides the characteristic polynomial of $t \in End(V)$.* □

THEOREM 6.14. *(Caley-Hamilton Theorem) Let t be an endomorphism on a vector space V of dimension n over a field F and let $f(x)$ be its characteristic polynomial in $F[x]$. Then $f(t)$ is the zero endomorphism on V.*

PROOF. We already know that $f(t)$ is an endomorphism on V, so

$$f(t)(0) = 0.$$

Now for a non-zero vector $w \in V$, we have seen that $C_w(t)$ is a t-invariant subspace of V with a basis $(w, t(w), ..., t^{m-1}(w))$ and the matrix of $t|_{C_w(t)}$ with respect to this basis is the companion matrix

$$A_g = \begin{bmatrix} 0 & 0 & \cdots & 0 & -g_0 \\ 1 & 0 & \cdots & 0 & -g_1 \\ 0 & 1 & \cdots & 0 & -g_2 \\ \vdots & \vdots & & \vdots & \vdots \\ 0 & 0 & \cdots & 0 & -g_{m-2} \\ 0 & 0 & \cdots & 1 & -g_{m-1} \end{bmatrix},$$

where $g(x) = g_0 + g_1 x + \cdots g_{m-1}x^{m-1} + x^m$ is the minimal polynomial of $t|_{C_w(t)}$ and furthermore, we also have seen that $g(x)$ characteristic polynomial of $t_{C_w(t)}$. Then by the above corollary, $f(x) = h(x) \cdot g(x)$ for some

$h(x) \in F[x]$. Thus, for any non-zero vector $w \in V$, $g(t)(w) = 0$ implies

$$f(t)(w) = 0.$$

□

COROLLARY 6.19. *(Matrix Version of Caley-Hamilton Theorem) Every square matrix A satisfies its own characteristic equation, i.e., if $\varphi(x) = \det(xI - A)$. then we have*

$$\varphi(A) = 0.$$

PROOF. Since we have an isomorphism

$$Mat_n(F) \approx End(F^n),$$

of F-vector spaces, each matrix A is an endomorphism on F^n, left multiplication by A. Thus, by the above theorem $f(A)$ is the zero endomorphism on F^n and hence it is the zero matrix, as claimed. □

Here is the second proof of the matrix version of Caley-Hamilton Theorem, which is more elementary than the above proof because we do not have to interpret a matrix as left multiplication on a finite dimensional vector space V.

THE SECOND PROOF OF COROLLARY 6.19. Write

$$\det(xI - A) = c_0 + \cdots + c_{n-1}x^{n-1} + x^n,$$

where $c_i \in F$. For our purpose, we denote this polynomial in x by

$$\varphi(x).$$

Thus,

$$\varphi(A) = \sum_{i=0}^{n} c_i A^i, \qquad c_n = 1,$$

is a polynomial in the matrix A. Then since we have

$$x^i I - A^i = (x^{i-1} I + \cdots + +xA^{i-2} + A^{i-1})(xI - A),$$

we can express A^i as

$$A^i = x^i I - (x^{i-1}I + \cdots + +xA^{i-2} + A^{i-1})(xI - A).$$

Thus,

$$\begin{aligned}\varphi(A) &= \sum_{i=0}^{n} c_i x^i I - \sum_{i=0}^{n} c_i(x^{i-1}I + \cdots + +xA^{i-2} + A^{i-1})(xI - A) \\ &= c(x)I - \sum_{i=0}^{n} c_i(x^{i-1}I + \cdots + +xA^{i-2} + A^{i-1})(xI - A),\end{aligned}$$

where $c(x)I = \det(xI - A)I$. On the other hand, we may recall that

$$c(x)I = adj(xI - A) \cdot (xI - A).$$

Thus, we have

$$\begin{aligned}\varphi(A) &= adj(xI - A) \cdot (xI - A) \\ &\quad - \sum_{i=0}^{n} c_i(x^{i-1}I + \cdots + A^{i-1})(xI - A) \\ &= \left\{ adj(xI - A) - \sum_{i=0}^{n} c_i(x^{i-1}I + \cdots + A^{i-1}) \right\} \cdot (xI - A) \\ &= D \cdot (xI - A),\end{aligned}$$

where D is a matrix with entries in $F[x]$. Then we note that the matrix $\varphi(A)$ on the left side of this equality does not contain the unknown x, whereas since the right side is the product of two matrices with entries in $F[x]$ where $xI - A$ is not the zero matrix, D must be the zero matrix. Otherwise, we write D as

$$D = D_k x^k + \ldots + D_1 x + D_0,$$

where D_k is not the zero matrix and each D_j, $0 \leq j \leq k$, is a matrix with entries only in F. Then the product, $D \cdot (xI - A)$ when expanded, has the non-zero leading term $D_k x^{k+1}$ in x, but this is impossible because $\varphi(A)$ on the left side is a matrix of constants (degree 0 in x). $\square$

COROLLARY 6.20. *The minimal polynomial of any* $n \times n$ *matrix divides the characteristic polynomial of* A*, and hence has degree at most* n.

PROOF. This is true by Theorem 6.13. □

REMARK 6.19. *Theorem 6.15 and Remark 6.21 below completely describe how we can obtain the minimal polynomial of an endomorphism on a finite-dimensional vector space* V *over a field* F.

6.2. A One to One Correspondence Between $F[t]$-modules and Endomorphisms on a Vector space V. Suppose that we are given an endomorphism on a vector space V, $T\colon V \longrightarrow V$ on a vector space V over a field F. We can use this endomorphism to make V into a module over the polynomial ring $F[t]$. For this

DEFINITION 6.25. *For a given polynomial* $f(t) = a_n t^n + \cdots + a_1 t + a_0 \in F[t]$ *and for a given vector* $v \in V$*, we define scalar multiplication:*

$$f(t).v = a_n T^n(v) + a_{n-1} T^{n-1}(v) + \cdots + a_1 T(v) + a_0 v.$$

We note that it is a generalization of left multiplication by a scalar $\alpha \in F$,

$$\alpha.v = \alpha v.$$

Then we can easily verify that

LEMMA 6.17. *Under this extended scalar multiplication, the module* V *over a field* F *becomes a module over the polynomial ring* $F[t]$ *in one unknown* t. □

Conversely, let V be an $F[t]$-module. Then by restricting scalars, we can easily check that V is a module over F. Moreover, we easily see that

LEMMA 6.18. *Left multiplication by the scalar* $t \in F[t]$,

$$t.v = tv$$

is an endomorphism T on V, $T\colon V \longrightarrow V$ sending $v \mapsto T(v) = tv$, and thus, left multiplication by a polynomial $f(t) = a_n t^n + \cdots + a_1 t + a_0 \in F[t]$,

$$\begin{aligned} f(t).v &= a_n t^n.v + a_{n-1}t^{n-1}.v \cdots + a_1 t.v + a_0.v \\ &= a_n t^{n-1}(T(v)) + a_{n-1}t^{n-2}(T(v)) \cdots + a_1 T(v) + a_0 v \\ &= a_n T^n(v) + \cdots + a_1 T(v) + a_0 v \end{aligned}$$

is an endomorphism on V. □

This establishes

PROPOSITION 6.11. *There exists a one to one correspondence between the set of all endomorphisms on a vector space V, $T\colon V \longrightarrow V$ over a field F, and the set of all modules V over the polynomial ring $F[t]$ in one unknown t.* □

That is,

REMARK 6.20. *An endomorphism on an F-vector space and an $F[t]$-module are equivalent concepts.*

6.3. Rational Canonical Form. Now if $\dim_F(V) = n$, then we can apply the theory developed in the last section of this Chapter: Given an endomorphism T on a finite-dimensional vector space V of dimension n over a field F, we view V as a *finitely generated* $F[t]$-module, and hence it has a *presentation matrix* A.

PROBLEM 1. *Think about how we are going to obtain a presentation matrix A of the finitely generated $F[t]$-module V from an endomorphism $t\colon V \longrightarrow V$. As the proof of the following theorem shows we are only able to obtain it very indirectly.*

THEOREM 6.15. *Let $t\colon V \longrightarrow V$ be an endomorphism of an n dimensional F-vector space V, and let $M = (M_{ij})$ be the matrix of t with respect to a basis $(v_1, ..., v_n)$ of V. Then the matrix*

$$M - tI \in F[t]$$

of polynomials in $F[t]$ represents the finitely generated $F[t]$-module V. That is, the matrix $M - tI$ presents the $F[t]$-module V.

PROOF. Since $(v_1, ..., v_n)$ still generates the $F[t]$-module V over $F[t]$, we have the canonical surjection

$$p\colon (F[t])^n \longrightarrow V.$$

Let $(b_1, ..., b_n)$ be a basis of the free $F[t]$-module $(F[t])^n$. By definition (cf. Theorem 6.1), each b_i is mapped to v_i under this p, that is,

$$p(b_i) = v_i, \qquad \text{for } i = 1, ..., n.$$

Then since F is a field it is a Noetherian ring, and hence $F[t]$ is a Noetherian ring, and hence $(F[t])^n$ is a Noetherian module by Proposition 6.5, and hence the kernel $\ker(p)$ is a finitely generated $F[t]$-module.

Claim. $\ker(p)$ is generated by n elements u_j of $(F[t])^n$,

$$u_j := b_j t - \sum_{i=1}^{n} b_i M_{ij} \in (F[t])^n, \qquad j = 1, 2, ..., n.$$

PROOF OF CLAIM. We note that $t(v_j) = \sum_{i=1}^{n} v_i M_{ij}$. Thus, for each $j = 1, 2, ..., n$, we have

$$\begin{aligned} p(b_j t - \sum_{i=1}^{n} b_i M_{ij}) &= p(b_j)t - \sum_{i=1}^{n} p(b_i) M_{ij} = v_j t - \sum_{i=1}^{n} v_i M_{ij} \\ &= t(v_j) - \sum_{i=1}^{n} v_i M_{ij} = 0. \end{aligned}$$

Thus, the submodule S of $(F[t])^n$ generated by these n elements $b_j t - \sum_{i=1}^{n} b_i M_{ij}$ over $F[t]$ is contained in $\ker(p)$. That is, $S \subseteq \ker(p)$. For the reverse inclusion, let

$$u := \sum_{j=1}^{n} b_j f_j(t)$$

be an element of $\ker(p)$. Then since each of its component b_jt^k can be recursively written as

$$\begin{aligned} b_jt^k &= b_jt \cdot t^{k-1} = (u_j + \sum_{i=1}^{n} b_iM_{ij}) \cdot t^{k-1} \\ &= u_jt^{k-1} + \sum_{i=1}^{n} b_it^{k-1}M_{ij} \\ &= u_jt^{k-1} + \sum_{i=1}^{n}(u_it^{k-2} + \sum_{\alpha=1}^{n} b_\alpha t^{k-2}M_{\alpha j})M_{ij} \\ &= \ldots\ldots. \\ &= u'_j + \sum_{i=1}^{n} b_i\gamma_{ij}, \end{aligned}$$

with $u'_j \in \ker(p)$ and $\gamma_{ij} \in F$, we can write u as an F-linear combination $\sum_{i=1}^{n} b_i\gamma_i$ of b_i's modulo S. Then since V is a vector space over F with basis $(v_1, v_2, ..., v_n)$ by hypothesis, $p(u) = 0$ forces all the scalars γ_i of b_i must be zero. Thus, $u \in S$. □

Thus, we *define* an $F[t]$-module homomorphism

$$s : (F[t])^n \longrightarrow (F[t])^n$$

by sending

$$b_j \longmapsto \sum_{i=1}^{n} b_iM_{ij} - b_jt = \sum_{i=1}^{n} b_i(M_{ij} - \delta_{ij}t), \qquad j = 1, 2, ..., n,$$

in order to have $img(s) = \ker(p)$. Moreover, this assignment explicitly tells us that the matrix of s with respect to the basis $(b_1, ..., b_n)$ of $(F[t])^n$ is $M - tI$. Therefore, we have

$$(F[t])^n / \ker(p) = (F[t])^n / img(s) = (F[t])^n / (M - tI)(F[t])^n$$

is the *cokernel* of s, and consequently $M - tI$ is a presentation matrix of the finitely generated $F[t]$-module V. □

REMARK 6.21. *Thus, applying the procedures explained in detail both in the proof of Theorem 6.9 and Proposition 6.3 to the matrix*

$$M - tI$$

we then obtain a "reduced" presentation matrix A of the $F[t]$-module V.

We recall that if the presentation matrix (with polynomial entries) A of an endomorphism $T\colon V \longrightarrow V$ is of size $r \times s$, then r is the number of chosen generators for the module V and s is the number of linear relations among these generators.

Regarding V as an $F[t]$-module, we can apply Theorem 6.12 to conclude that V is a direct sum of cyclic $F[t]$-submodules W_i,

$$V = W_1 \bigoplus \cdots \bigoplus W_k,$$

where each W_i is isomorphic to $F[t]/(p_i^{e_i})$, each $p_i(t)$ being an irreducible polynomial in $F[t]$. Since we are assuming that V is a finite-dimensional vector space, we have no free summand in this expression.

DEFINITION 6.26. *Such a module is called a* ***torsion module of finite type****.*

In fact, we have seen that Theorem 6.15 and Remark 6.21 tell us how to compute each invariant factor, $p_i^{e_i}$, from the matrix $M - tI$, where M is the matrix of the endomorphism $T\colon V \longrightarrow V$.

DEFINITION 6.27. *The last polynomial $p_k^{e_k}$ (not in part b but) in part a of Theorem 6.12 is called the* ***minimal polynomial*** *of T. Or equivalently, the least common multiple of those invariant factors $p_i^{e_i}$ in part b of Theorem 6.12 is called the* ***minimal polynomial*** *of T.*

Now, we definitely need to translate such a direct sum into an information on an endomorphism $T\colon V \longrightarrow V$ defined by

$$T(v) = tv.$$

For this, we first recall Corollary 6.15 which says that a cyclic $F[t]$-module W is isomorphic to the quotient $F[t]$-module

$$F[t]/(f(t)),$$

where

$$f(t) = t^m + a_{m-1}t^{m-1} + \cdots + a_1 t + a_0$$

is a monic polynomial in $F[t]$. This is because $F[t]$ is a *PID* (principal ideal domain). Thus, if we denote w_0 be the residue of $1 \in F[t]$ in $F[t]/(f(t))$, i.e., $w_0 = 1 + (f(t)) \in F[t]/(f(t))$, we have

$$f(t)w_0 = 0$$

in $F[t]/(f(t))$. And by this construction, we see that $f(t)$ generates the module of relations, that is, the set $\{f(t)\}$ is a complete set of relations.

On the other hand, we note that the elements $w_0, tw_0, ..., t^{m-1}w_0$ form an F-basis for $F[t]/(f(t))$, because the set $\{t^i : i = 0, 1, 2, ...\}$ is an F-basis of the F-vector space $F[t]$. If we denote this basis by $w_i = t^i w_0$, we then have

$$tw_0 = w_1, tw_1 = w_2, ..., tw_{m-2} = w_{m-1}.$$

In addition to this, since we have

$$\begin{aligned} 0 &= f(t)w_0 \\ &= (t^m + a_{m-1}t^{m-1} + \cdots + a_1 t + a_0)w_0 \\ &= tw_{m-1} + a_{m-1}w_{m-1} + \cdots + a_1 w_1 + a_0 w_0, \end{aligned}$$

we see that the action of t on w_{n-1}:

$$tw_{m-1} = -a_{m-1}w_{m-1} - \cdots - a_1 w_1 - a_0 w_0.$$

Therefore, since T acts as left multiplication by t, we have

$$T(w_0) = w_1, T(w_1) = w_2, ..., T(w_{m-2}) = w_{m-1},$$

and

$$T(w_{m-1}) = -a_{m-1}w_{m-1} - \cdots - a_1 w_1 - a_0 w_0.$$

This exactly means that the matrix of T w.r.t the basis $(w_0, w_1, ..., w_{m-1})$ is

$$T(w_0, w_1, ..., w_{m-1}) = (w_0, w_1, ..., w_{m-1})M_f,$$

where M_f is the companion matrix of the polynomial

$$f(t) = t^m + a_{m-1}t^{m-1} + \cdots + a_1 t + a_0 :$$

$$M_f = \begin{bmatrix} 0 & 0 & \cdots & 0 & -a_0 \\ 1 & 0 & \cdots & 0 & -a_1 \\ 0 & 1 & \cdots & 0 & -a_2 \\ \vdots & \vdots & & \vdots & \vdots \\ 0 & 0 & \cdots & 0 & -a_{m-2} \\ 0 & 0 & \cdots & 1 & -a_{m-1} \end{bmatrix}.$$

REMARK 6.22. *We note that the characteristic polynomial* $\det(M_f - tI)$ *of* M_f *is equal to* $f(t)$.

Secondly, we note that each $F[t]$-submodule W_i is a T-invariant F-subspace of the F-vector space V. Thus, if we choose a basis $\mathbf{B}_i$ for each T-invariant subspace W_i and put them together to have a basis $\mathbf{B} = (\mathbf{B}_1, ..., \mathbf{B}_k)$ of V, then the matrix of T looks something like an identity matrix but with block form:

$$\begin{bmatrix} M_1 & & & & \\ & M_2 & & & \\ & & \cdot & & \\ & & & \cdot & \\ & & & & M_k \end{bmatrix},$$

where each M_i is the matrix of the restriction of T to W_i. This proves

THEOREM 6.16. *Let T be an endomorphism on a finite-dimensional vector space V over a field F. Then there exists a basis for V with respect* *to which the matrix of T consists of blocks of the type:*

$$[-a_0], \begin{bmatrix} 0 & -a_0 \\ 1 & -a_1 \end{bmatrix}, \ldots, \begin{bmatrix} 0 & 0 & .. & 0 & -a_0 \\ 1 & 0 & .. & 0 & -a_1 \\ 0 & 1 & .. & 0 & . \\ . & . & .. & . & . \\ . & . & & . & . \\ 0 & 0 & .. & 0 & -a_{m-2} \\ 0 & 0 & .. & 1 & -a_{m-1} \end{bmatrix}.$$

COROLLARY 6.21. *The characteristic polynomial* $\det(T - tI)$ *of T is the product of the polynomials $f_i(t)$, called the **invariant factors** of T, each of which is associated to the companion matrix M_{f_i}'s.* □

REMARK 6.23. *We note that the invariant factors of T is exactly the same as those d_i's (not in part b but) in part a of Theorem 6.12. Thus, the problem is all about how to compute these invariant factors, and by Theorem 6.15 we know that they are computed from the matrix*

$$M - tI,$$

where M is the matrix of T with respect to some basis of V. Then from Theorem 6.16 and part a of Theorem 6.12 we determine the list *of the invariant factors of the endomorphism T. Moreover, Remark 6.18 says that this list is uniquely determined.*

DEFINITION 6.28. *Such a form in this theorem for the matrix of a linear operator is called a **rational canonical form**.*

We note that these blocks are located along the diagonal entries of the identity matrix, and it is the best form available over an arbitrary field, namely, over a field which is not algebraically closed.

The following example tells us how one can obtain such a form from a given matrix for a linear operator T.

EXAMPLE 6.19. *As a continuation of Example 6.7, we have seen Example 6.17. Since*

$$V \approx F[t]/(g(t)) \approx F[t]/(t-1)^2 \times F[t]/(t-2),$$

and since $(t-1)^2 = t^2 - 2t + 1$, *the rational canonical form of the matrix of the linear operator on* V *in Example 6.7 is:*

$$\left[\begin{array}{cc|c} 0 & -1 & \\ 1 & 2 & \\ \hline & & 2 \end{array}\right],$$

where the matrix in the first block is the companion matrix of $(t-1)^2$ *and the matrix in the second block is the companion matrix of* $t-2$.

Here, we note that the invariant factor of V *is* $g(t)$ *by part a of Theorem 6.12.*

EXAMPLE 6.20. *(MAPLE) The following lines compute the rational canonical form of a given matrix:*

```
———————————
interface(echo=2);
with(linalg);
a:=matrix([[0,2,0,-6,2],[1,-2,0,0,2],[1,0,1,-3,2], \
[1,-2,1,-1,2],[1,-4,3,-3,4]]);
reduct1:=ratform(a,'Q');
print(Q);
evalm(Q^(-1)&*a&*Q);
minpoly(a,x);
charpoly(a,x);
simplify(charpoly(a,x)/minpoly(a,x));
———————————
```

6.4. Jordan form. In this subsection, we further specialize our discussion when the bottom field F is the field $\mathbb{C}$ of complex numbers.

Everybody knows that irreducible polynomials in $\mathbb{C}[t]$ are all linear, $p(t) = t - \alpha$ for some $\alpha \in \mathbb{C}$, thus, by part a of Theorem 6.12, every finitely generated $\mathbb{C}[t]$-module V is a direct sum of submodules which are isomorphic to ones of the form

$$\mathbb{C}[t]/(t-\alpha)^n.$$

This is a vector space over $\mathbb{C}$. As before we let w_0 be the residue of 1 in $\mathbb{C}[t]/(t-\alpha)^n$, i.e., $w_0 = 1 + (t-\alpha)^n$.

Thus identifying V, for instance, with $\mathbb{C}[t]/(t-\alpha)^n$, we consider the image of w_0 under the powers $(T-\alpha)^i$ of the linear operator,

$$(T-\alpha) : V \longrightarrow V,$$

$$\begin{aligned} w_1 &= (t-\alpha)w_0, \\ w_2 &= (t-\alpha)^2 w_0, \\ &\cdots \\ w_{n-1} &= (t-\alpha)^{n-1} w_0. \end{aligned}$$

By a similar argument we have done before, we see that the vectors

$$w_0, w_1, ..., w_{n-1}$$

are linearly independent over $\mathbb{C}$, but

$$w_n = (t-\alpha)^n w_0 = 0$$

in $\mathbb{C}[t]/(t-\alpha)^n$. Thus, we have

$$\begin{aligned} Tw_i &= w_{i+1} + \alpha w_i, \qquad i = 0, ..., n-2, \\ Tw_{n-1} &= \alpha w_{n-1}. \end{aligned}$$

This establishes

THEOREM 6.17. *Let $T\colon V \longrightarrow V$ be an endomorphism on a finite-dimensional complex vector space V. Then there exists a basis of V with* *respect to which the matrix of T consists of blocks of the type:*

$$[\alpha], \begin{bmatrix} \alpha & 0 \\ 1 & \alpha \end{bmatrix}, \begin{bmatrix} \alpha & 0 & \cdots & 0 & 0 \\ 1 & \alpha & \cdots & 0 & 0 \\ 0 & 1 & \cdots & 0 & 0 \\ \vdots & \vdots & & \vdots & \vdots \\ 0 & 0 & \cdots & \alpha & 0 \\ 0 & 0 & \cdots & 1 & \alpha \end{bmatrix}.$$

REMARK 6.24. *Explanation on how to obtain these blocks, or, equivalently, the list of invariant factors, is given in Remark 6.21 and Remark 6.23.*

DEFINITION 6.29. *Such a matrix in this theorem is called a **Jordan form**, or a **Jordan matrix**.*

Since a Jordan matrix J is lower triangular, we easily see that its eigenvalue is α, and its associated eigenspace

$$E_a = \{v \in V : JX = \alpha X\}$$

is generated by a single eigenvector. In this subsection we thus have established:

COROLLARY 6.22. *Given a complex matrix A there exists an invertible complex matrix P such that PAP^{-1} is a Jordan matrix, which is unique up to permutation of the blocks.* □

EXAMPLE 6.21. *The Jordan form or Jordan matrix of the module in Example 6.17 consists of two Jordan blocks:*

$$\left[\begin{array}{cc|c} 0 & 1 & \\ 1 & 1 & \\ \hline & & 2 \end{array}\right].$$

We note that computing the Jordan form of a given matrix requires finding eigenvalues, i.e., roots of its characteristic polynomial $p(t)$:

$$p(t) = \det(tI - A) = 0.$$

If the eigenvalues are all distinct, the Jordan form is the diagonal matrix:

$$\begin{bmatrix} \alpha_1 & & & & \\ & \cdot & & & \\ & & \cdot & & \\ & & & \cdot & \\ & & & & \alpha_n \end{bmatrix}.$$

Otherwise, let $\alpha_1 = \alpha$ be an r-multiple root of the characteristic equation: $p(t) = 0$. Then depending on $r = 1, 2, 3, ...$, there are various possibilities for the part of the Jordan matrix with diagonal entries α.

They can be distinguished by computing eigenvectors of various operators

$$(\alpha I - A)^i$$

related to T.

REMARK 6.25. *This is about* ***generalized eigenvectors*** *of an endomorphism A.*

In fact, it can be shown that

PROPOSITION 6.12. *The dimension of the null spaces of the operators*

$$(\alpha I - A)^i, \qquad i = 1, 2, ..., r/2,$$

distinguishes the Jordan forms in all cases. □

EXAMPLE 6.22. *When $r = 4$, we have in total 5 different possibilities, basically depending on the number of distinct eigenvectors:*

$$(i) \begin{bmatrix} \alpha & & & \\ 1 & \alpha & & \\ & 1 & \alpha & \\ & & 1 & \alpha \end{bmatrix}, (ii) \left[\begin{array}{ccc|c} \alpha & & & \\ 1 & \alpha & & \\ & 1 & \alpha & \\ \hline & & & \alpha \end{array}\right], (iii) \left[\begin{array}{cc|cc} \alpha & & & \\ 1 & \alpha & & \\ \hline & & \alpha & \\ & & 1 & \alpha \end{array}\right],$$

$$(iv) \left[\begin{array}{cc|c|c} \alpha & & & \\ 1 & \alpha & & \\ \hline & & \alpha & \\ \hline & & & \alpha \end{array}\right], (v) \left[\begin{array}{c|c|c|c} \alpha & & & \\ \hline & \alpha & & \\ \hline & & \alpha & \\ \hline & & & \alpha \end{array}\right].$$

Thus, the dimensions of eigenspaces are $1, 2, 2, 3,$ *and* 4. *Thus this dimension distinguishes all cases except the second and the third cases. But, these remaining two cases can be distinguished by the matrix* $(\alpha I - A)^2$. *It is zero in the third case and not zero in the second case.*

EXAMPLE 6.23. *(MAPLE) The following lines compute the Jordan form of a given matrix:*

———————

```
interface(echo=2);
with(linalg);
a:=matrix([[1,-1,1],[1,-1,1],[1,-1,0]]);
reduct1:=jordan(a,'P');
print(P);
evalm(P^(-1)&*a&*P);
minpoly(a,x);
charpoly(a,x);
```

———————

Exercises for Section 6.6

EXERCISE 6.24. *Determine the Jordan form of the matrix*

$$A = \begin{bmatrix} 1 & 1 & 0 \\ 0 & 1 & 0 \\ 0 & 1 & 1 \end{bmatrix}.$$

EXERCISE 6.25. *Determine the Jordan form of the matrix*

$$A = \begin{bmatrix} 1 & 1 & 1 \\ -1 & -1 & -1 \\ 1 & 1 & 1 \end{bmatrix}.$$

APPENDIX A

An Introduction to Algebraic Analysis

1. Tools in Mathematics

The whole body of Mathematics consists of many different kinds of knowledge, or results and procedures which produce such results, most of which are verified through the process of the so-called "proof". Those proved and somewhat useful or important are usually called **theorems**. Unproven ones are formally called **conjectures**. The place for most mathematicians should be between these two words, because we are born to establish reasonable conjectures and then prove them. When one tries to prove something, something itself must be first recognized as a reasonable mathematical object; discovering such object could be quite personal.

In proving reasonable conjectures, one generally uses two tools, a special type of logics, called **mathematical logic**, and **computation,** definitely guided by his own mathematical intuitions or experiences. Since they are applied as tools in the process of the so-called **proof**, they themselves cannot be called mathematics. However, the kinds of the methods employed certainly put some unavoidable limitations on the character of the consequences. These two tools, nevertheless, do not hurt each other. In contrast, they help each other, and these two are the most powerful tools we, mathematicians, have. Depending on the degree of putting more emphasis on any of these two, one usually has two basic tendencies in doing mathematics, axiomatic or computational.

1.1. Axiomatic Nature. Although it is very difficult to figure out what constitutes mathematics, one may describe its axiomatic nature as follows: Once a reasonable mathematical conjecture is recognized, one accepts, without questioning their existence or validity, a finite number of undefined terminologies and that of postulates (i.e., axioms or principles) which describe some relations between the undefined terminologies, and one applies a finite number of steps of mathematical logic directly to them or to some of their consequences until one finally reaches the desired positive or negative conclusions. As a demonstrative science started even before Greek period, mathematics has shown this axiomatic nature for a long time. One can find such an evidence in, for instance, Euclid's Elements [**E**] and other works of Greek mathematicians and philosophers. In fact, one can even trace back to their works to see that most of the formal terminologies used in mathematics, something like definitions, propositions, corollaries, etc., were already used in his Elements and other Greek literatures. The meaning and usage of these terminologies and other metamathematical concerns are also well explained in the introduction of A. Whitehead and B. Russell's famous work, *Principia Mathematica* [**WR**], which was originally published in 1910.

1.2. Computational Nature. When computations in general are concerned, we know that there are basically two different kinds, namely, exact computation and approximation. **Approximation** is done through the limiting process mostly in terms of inequalities, where as **exact computation or symbolic computation** is done through the so-called four binary operations: addition, subtraction, multiplication and acceptable division. These operations are applied not only to numbers but also to a mixture of numbers and **symbols**.

Computations are very important to us at least for the following three reasons. First, they usually show up in the middle of the process called proof. This is almost unavoidable to every mathematician. Secondly, a decent amount of computational experiments quite often reveals to us what's going on in a very clean manner, while simple-minded 'what if ...' ways of logical questioning cannot. Thirdly, we, mathematicians, can communicate with other branches of sciences very efficiently and successfully, through computational results and their usefulness. Considering that mathematics has so far grown up to this much within the accumulation of our civilization, this point of view is also very important, because isolation only hurts oneself till one's starvation.

Up to the end of the 19th century or the beginning of the 20th century, some important mathematical works discovered and proved were based on individual mathematician's heavy hand-computational experiments. However, as was the case in Invariant Theory of the 19th century, mathematicians were practically forced out to give up heavy computations when the number of unknowns and the size of numbers are substantially increased so that hand-calculation obviously becomes terrible. Any attempt based on computational experiments may be called a **computational approach**.

2. Overemphasis on Axiomatic Approach in the 20th Century

The revival of the so-called **axiomatic approach** after the Greek period (namely, Euclid's Elements) is clearly evident in the Lecture Notes of E. Artin and E. Noether, written up by B.L. van der Waerden in 1930s [**BW**]. This approach, especially well-supported by precise notations adapted from the subject, Set Theory, was so good in pursuing a particular-minded concept or property, that it rapidly pushed out the old tradition of computational approaches, which was quite common up to the first quarter of 20th century, from the mathematical world. In addition, some eminent mathematicians, like Banach in Analysis or later E. Spanier in Algebraic Topology [**ES**], tried to figure out and establish sufficiently small number of axioms but still good enough to explain successfully the rest of the theory they know. In order to fulfill such tasks, they examined many computed examples of different kinds and pulled out some common underlying axioms which replace too much involved terrible computations. The success of this attempt is quite evident in almost every part of current mathematics in both research articles and textbooks. It is also around this period that most mathematicians officially but still reluctantly began to acknowledge the usefulness of the non-traditional mathematical axiom, i.e., the axiom of choice, or its equivalence, the Zorn's Lemma.

The point of views expressed in this article could be quite personal until they get more acceptance from general audience. However, the readers will not be hurt by the author's viewpoints on mathematics expressed in this article but it will supply a fresh taste or sense to our main meal, mathematics.

2.1. Danger of Too Much Emphasis on Axiomatic Approach. One should note that those mathematicians, who introduced an appropriate set of axioms for the knowledge they already know, were also very good at computation, since they were trained as, called from now on, the "first" generation mathematicians, by their teaches most of whom were under the strong influence of computational approach of the 19th century. On the other hand, the "second" generation mathematicians, i.e., those who were mathematically trained by the first generation mathematicians, tended to start their mathematics not from heavy computational experiments but from a small amount of axioms, and they were more and more away from computational experiences. The lack of computational experiences is more significant in the "third" generation mathematicians, i.e., those trained by the second generation mathematicians.

Nevertheless, the strength of axiomatic approach had been fully demonstrated by the monumental works of a single mathematician, Alexander Grothendieck (1928 -), on the Foundation of Algebraic Geometry, [**AG1**], FGA [**AG2**], EGA's [**AG3**] and [**AG4**], and SGA's [**AG5**] during 1950s up until the beginning of 1970s. His works in turn simply recall the works of a group of mathematicians, called Bourbaki [**B2**] and [**B1**], as elementary results. Anyone who is very much interested in the scope of axiomatic approach in mathematics explored during the 20th century is strongly encouraged to study and digest the above mentioned Grothendieck's works, of which the basic concepts and results are described in terms of categorical languages, called the **language of schemes**.

In particular, the tendency of the results obtained under a strong influence of axiomatic approach is too much conceptual, and hence becomes more and more **soft** and hence fragile, as time goes. In this case, a good-minded mathematician should worry about whether his and/or her results will still survive after, say, $1,000$ years from now. A prototypical example is Euclid's algorithm for computing the greatest common divisor of two numbers. His method is still known to us because it is about computation of certain mathematical objects in an efficient manner. Moreover, lack of computability only speeds up the isolation of Mathematics from other branches of Sciences. This is in no way acceptable to us.

For an elegant axiomatic treatment but still with a strong emphasis on computational point of views, of Algebraic Geometry, one may refer to Prof. George R. Kempf's two books [**GK1**] and [**GK2**].

3. A New Computational Tool: Computers

3.1. Personal Computers and High Performance Servers. Starting at around the beginning of the fourth quarter of the 20th century, computer industry has grown up with an unexpected rate of success; this nowadays allows us to compute very big numbers using a Personal Computer or even a Notebook Computer. Recently, inexpensive High Performance Cluster Servers are also available to us. As a tool for actual computation, computers are already playing a significant role and such role will surely change the mathematical world of the 21-st century. Here are some examples.

EXAMPLE A.1. *Here, $p(n)$ denotes, by definition, the number of ways of writing a given natural number n as a sum of others (called components) ignoring the order among the components. One starts with a very famous theorem:*

THEOREM A.1. *(Rademacher* **[GA]***) Given a natural number n, one has*

$$p(n) = \frac{1}{\pi\sqrt{2}} \sum_{k=1}^{\infty} A_k(n)\sqrt{k}\frac{d}{dn}\frac{\sinh((\pi/k)\sqrt{(2/3)(n-\frac{1}{24})})}{\sqrt{n-\frac{1}{24}}}, \tag{A.1}$$

where

$$A_k(n) = \sum_{\substack{0 \le h < k \\ (h,k)=1}} \varpi_{h,k} \exp(-2\pi i n h/k),$$

where

$$\varpi_{h,k} = \exp(\pi i s(h,k)),$$

is a $24k$th root of unity and where

$$s(h,k) = \sum_{\mu=1}^{k-1} \left(\frac{\mu}{k} - \left[\frac{\mu}{k}\right] - \frac{1}{2}\right)\left(\frac{h\mu}{k} - \left[\frac{h\mu}{k}\right] - \frac{1}{2}\right)$$

is the so-called Dedekind sum.

It turns out that the value of $p(20,000,000)$, when computed it using this theorem, requires more than $4,900$ number of decimal digits (Cf. **[STS1]***). Thus the final result is really a super-ultra big number, which is one point. Moreover, another point is that it only took about 17 Minutes with already-done pre-calculations of the so-called Dedekind sums, under the same hardware and software limitations as in the next example. These two points clearly show us the computing capability of current PC's.*

REMARK A.1. *One should note that computation of such a super-ultra number eventually raises a serious concern whether the computed result is absolutely correct. This is certainly a new kind of headaches to mathematicians.*

On the other hand, the author recently discovered an exact algebraic formula for a graded partition function $p_{\mathbf{M}}(n)$ over a multiset $\mathbf{M}$ and an associated quasi-recursive formula for $p_{\mathbf{M}}(n)$, both of which never *require approximation at all (Cf.* **[STS2]***). One should note that it is neither quite* *possible to obtain an analytically exact formula for* $p_{\mathbf{M}}(n)$ *nor to obtain a recursive formula for it, something like either Rademacher's formula or the well-known Euler's recursive formula for* $p(n)$*, unless there is a significant symmetry among the elements of* $\mathbf{M}$*. However, one version of the author's formulas computes all the values of* $p_{\mathbf{M}}(i)$ *for* $1 \leq i \leq n$ *with time efficiency of* $O(n^2)$ *for a general multiset* $\mathbf{M}$ *(Cf.* [**STS3**]*).*

3.2. Computer Algebra Systems. Of course, having just a very fast computing machine of which hardware is very reliable is not enough. One also needs a reliable software which guarantees several facilities. Among them, the most urgent requirements are, for instance, ability to handle numbers of any sizes and arbitrary precision in representing fractions, each of which absolute reliability is definitely required in any computation with computers.

EXAMPLE A.2. *The following Table shows the actual computation of the value of the classical partition function* $p(n)$*, for* $n = 200$*, using the Rademacher's formula in Eq (A.1). Here, by adding the first* 8 ***non-zero*** *terms of fractional real numbers, one gets a fractional real number,*

$$3972999029388.001346332,$$

which is very close to the true integer value $p(200) = 3972999029388$.

$$\begin{array}{rcr} t(1) & = & +3.97299899318589575787 2E+12 \\ t(2) & = & +36282.9784854411847056 \\ t(3) & = & -87.58391287399 \\ t(4) & = & +5.146534584 \\ t(5) & = & +1.424217569 \\ t(6) & = & +0.07102143214 \\ t(7) & = & 0 \\ t(8) & = & +0.04357655366 \\ t(9) & = & +0.0256657539 \\ [t(10) & = & +0.008603641042] \\ \hline \text{Up to 9th Partial Sum} & = & +3972999029388.001346332 \end{array}$$

When it is about computation of an exact integer value in this way of approximation, there should not be any approximation-related errors in all of the intermediate steps. Do not expect that an error-free computation of such a kind is quite easy (Cf. [**STS1**]*), unless one uses very reliable Computer Algebra Systems.*

REMARK A.2. *This example, which is prepared by the author of this book, is especially interesting because it is a direct computational verification of the famous example computed by Hardy and Ramanujan (p. 284 of* [**HR**]*) and quoted by Andrews (p. 70 of* [**GA**]*). Indeed, the original calculation in Hardy's own handwriting occurs on the dust jacket of the 1969 edition of "A Mathematician's Apology",* [**H**]*. To our surprise, the calculation by Hardy and Ramanujan contains an error. The corrected calculation as in Example A.2 turns out to be much closer to* $p(200)$ *than the original incorrect version. Therefore, the third term,* -87.555*, in the original computation table for* $p(200)$ *in p. 284 of* [**HR**] *and later quoted in p. 70 of* [**GA**]*, needs to be replaced with* -87.584*.*

Any software package lacking these two capabilities should not be our choice for a tool for computational experiments. In addition to them, it is also required to have an ability to solve a given system of polynomial equations, i.e., non-linear equations over, say, the field of complex numbers. This means that it must have an ability to handle symbols as well as numbers. The theory for such computation had not fully understood until Bruno Buchberger (1942 -) discovered his famous algorithm for Groebner bases in 1965. Any reliable software package with at least these three facilities could be called a **Computer Algebra System**.

There are two types of Computer Algebra Systems, commercially available or not. Some of the commercially available ones are, MAPLE, MACSYMA, REDUCE, etc. Some of non-commercially available ones are, MACAULAY 2, CoCoA, etc. There should be more than these.

3.3. Behind Computational Experiments. One first notes that there are two kinds of computations one can do with computers, **exact computation** and **approximation**. Exact computation only needs addition, subtraction, multiplication, and allowed division depending on given conditions, on a given mixture of known symbols, unknown symbols, and numbers. In contrast to this, approximation allows us to apply non-exact division and limiting process, as well as addition, subtraction, multiplication and division on given numbers only. Thus, exact computation allows us to manipulate not only numbers but also **symbols**, whereas approximation only allows us to manipulate numbers. This tells us why exact computation is also called **Algebraic Analysis** or **Symbolic Analysis**, while approximation is called **Numerical Analysis**. Examples of Numeric Analysis are numerous. For instance, FEM (Finite Element Method) is just one of them. Also, see Example A.2 above.

These two approaches in computational methods do not hurt each other. In fact, they help each other. The only general trouble is that Algebraic Analysis does require some amount of elementary Algebra, mostly about polynomial rings in more than one unknowns, as well as some amount of Modern Algebraic Geometry, developed by Oscar Zariski [**OZ**] and J. P. Serre [**JS**], and a few others (For beginners, this amount of accounts on the subject may be good enough).

A proper sequence of computational approach is as follows:

(1) to apply appropriate exact methods in Algebraic Analysis until all available ones are exhausted, and then, as a final step,
(2) to apply appropriate approximation methods in Numeric Analysis.

The subject of Numeric Analysis has been well known to us after Sir Isaac Newton. However, Algebraic Analysis is very new to us. The most important theoretical background for recent revival of computational approach is the Buchberger's astonishing discovery of the algorithm for Groebner bases for polynomial ideals in more than one indeterminates. He figured out the method while he was working on his Ph.D. thesis, and it was first recognized by mathematicians oriented to computer science somewhat before or around 1975. I myself learned this remarkable algorithm in mid 1980s.

3.4. Danger of Too Much Emphasis on Computational Approach. If one puts too much emphasis on computational approach, he will soon find that he does not know why the method works and eventually he will get lost, because the arithmetic operations themselves never tell him what's going on, unless he knows *in advance* what's really going on. Unfortunately, it is obviously impossible most of the time, especially when one works on some mathematical phenomena not yet well understood to him. This means that one really needs a well-developed theory for exact computation or Algebraic Analysis.

There was no general theory for exact computation with computers until B. Buchberger discovered a very important algorithm. The mixture of his algorithm with the division algorithm produces very important algorithms for polynomial ideals in more than one variables. It is very urgent to understand this theory for hard but honest and diligent computation. The readers of this article should not underestimate the strength and usefulness of this theory.

4. Expected Trends of the 21st Century Mathematics

4.1. Computer Age. It is well known that there were two types of computers, analogue and digital, in the past. Nowadays, however, computers usually mean **digital computers** or their variants for better performance, since performance of digital computers (for instance, signal to noise ratio) has already proved to be a lot better than analogue types. It is also well known that the theoretical creation of digital computers was done by a famous mathematical logician, von Neumann, who invented Class Theory to answer, for instance, the question like whether the set of all sets is a set or not. After theoretically creating digital computers, he immediately developed a good deal of approximation theory for computing the values of non-polynomial functions with his theoretical machines.

Although Electric Engineering, Computer Engineering, and Computer Science are respectively needed to establish hardware, architecture, and operating system and software, the final combination of them is a system, called, computer, and the main reason for developing such a system is to compute or calculate something in any possible senses. The point we, mathematicians, should not miss is that **computation belongs to mathematics**. Thanks to computers, paperless computation of huge numbers is already available to those who know how to use some of the commercially or non-commercially available Computer Algebra Systems, so that the scope of Algebra and Algebraic Geometry in both pure and applied directions has already been expanded a lot.

4.2. Computers Are for Computation Related Human Activities. Since computation solely belongs to mathematics, it is mathematician's due obligation to examine and develop the needed mathematical algorithms in order to obtain from the machine the desired results with absolute correctness. In addition to this, our second obligation is to make it as efficient as possible. If it takes too much computation time, then such a computational approach will never buy other people's approval at all. To my regret, these two can't be easily accomplished by those who were not trained as mathematicians.

Unlike any other centuries, existence of so many mathematical objects have been proved during the 20th century, mostly thanks to active application of axiomatic methods influenced by Set Theory or Category Theory. These results are all inputs for further investigation that which of them are computationally constructible, applying the methods of Algebraic Analysis.

I should mention one dangerous aspect of the Zorn's Lemma or its equivalent axioms. If one used the Zorn's Lemma in some place of his proof of the existence of certain mathematical object, then his proof is computationally useless in the sense that it never really tells us how to construct such an object only using the four arithmetic operations and the limiting process. This is really a disappointing aspect of the Zorn's Lemma, but is, as long as **algorithmic construction** is concerned, still important enough to keep in mind in order for us not to use it in our proof as possible as we could.

Computers have already become a convenient and very capable tool for computations in any part of mathematics. We do not necessarily require to have an access to Supercomputers, Servers, or even Workstations for computational experiments. We only need to have Personal Computers for most computational experiments. Since Personal Computers are nowadays available to everyone and since they are already too fast and too capable of, most computational experiments could be done with them without spending too much time and money. Moreover, the hardware-related performance of these machines are nowadays very much reliable. The author strongly believes that such a very reliable computational environment, already available to everybody, sooner or later changes the whole atmosphere of mathematics of the 21-st century. For instance, the so-called **internet computing** will soon become very common.

One of the main ingredients of mathematics is perhaps its **exactness** in the process of both **proof** and **computation.** The point is that both of them require quite different types of exactness. This may be also the main reason why no other branches of Sciences even try to compete with mathematics, but only to rely on the results produced by mathematicians. Consequently mathematics has its own everlasting life. We should not give up any of these two powerful tools, proof and computation, to other branches of Sciences. On the other hand, we have been silently forced to ignore the significance of computational aspects for at least a century, mostly because of our inability in exactly hand-computing mixtures of big numbers and many parameters and also partly because of strong influences of axiomatic approach of the 20th century. Isn't it a time for us to revive computation related mathematical activities with a new tool, computers, to balance ourselves between them?

5. About Our InetCompu Service

We have recently started an E-mail based Internet Computing Service, called InetCompu.

If one visits

http://trinitas.mju.ac.kr/intro2InetCompu.html

the detailed information and several demonstrative services are provided:[1]

5.1. Our Concept and Terminology. If a person, for instance, using PC, PDA, or Mobile Phone, can perform the kinds of computations he wants on a remote Server, where he is not even required to have his own account, and is also allowed to get back the computed results immediately, then we call such a service, **InetCompu** (Internet Computing, Copyrighted) or **GridCompu** (Grid Computing, Copyrighted).

5.2. What We Have Done So Far. Several years ago, the author of this book figured out a practical method to actually realize the needed Software Interface (called General Purpose Grid Middleware Interface) for such an InetCompu Service, and he finished the development of the first primitive but working version of the Interface at around August 16, 2001. Later on, the Interface for InetCompu was significantly upgraded by him to such a degree that it could handle the in-comming data more neatly and smartly. The final version currently in usage also passed a rigorous local fields test successfully.

As of August 28, 2001, we began to provide an InetCompu Service, i.e., whoever sends an E-mail to a particular E-mail address of our Server, for example, containing

(1) either an input data for an application program in our Server for a certain mathematical formula (for instance, input: n=1234567 to numbpart@trinitas.mju.ac.kr in plain text format

(2) or his own program, for instance, written in Reduce commands,

then he will soon receive the computed results from our Server.

[1] http://ch.catholic.or.kr/pundang/archive/4/intro2InetCompu.html

5.3. An Example of InetCompu.

In particular,

http://trinitas.mju.ac.kr/intro2numbpart.html

explains how one can use our InetCompu Service to compute the value of $p(n)$ for a given input n, which utilizes Rademacher's formula for $p(n)$ in Theorem A.1.[2]

[2]http://ch.catholic.or.kr/pundang/archive/4/intro2numbpart.html

Bibliography

[AL] W.W. Adams and P. Loustaunau, *An Introduction to Gröbner Bases,* GSM Vol. 3, AMS, 1994.

[GA] G.E. Andrews, *The Theory of Partitions,* Reading, 2nd printing,1981, Addison-Wesley

[BW] T. Becker and V. Weispfenning, *Groebner Bases: A Computational Approach to Commutative Algebra*, Springer Verlag, 1993.

[B1] N. Bourbaki, *Algebra,* Part I and II (in English), Springer-Verlag, 1970, 1981.

[B2] N. Bourbaki, *Commutative Algebra* (in English), Springer-Verlag, 1985.

[CLO] D. Cox, J. Little, and D. O'Shea, *Ideals, Varieties, and Algorithms, An Introduction to Computational Algebraic Geometry and Commutative Algebra,* 2nd Ed., Springer-Verlag, 1997.

[ER] D. Eisenbud and L. Robbiano, *Computational Algebraic Geometry and Commutative Algebra,* Symposia Mathematica Vol. XXXIV, Istituto Nazionale di Alta Mathematica Francesco Severi, Cambridge Univ. Press, 1993.

[DE] D. Eisenbud, *Commutative Algebra with a View Toward Algebraic Geometry,* Springer-Verlag, 1995.

[E] Euclid, *The Elements,* Vol. 1 - 3 (English translation), 2nd Ed., Dover, 1956.

[AG1] A. Grothendieck, *Algébre Homologique,* Secrétariat mathématique, 11 rue Pierre Curie, PARIS 5e, 1958.

[AG2] A. Grothendieck, *Fondements de la Géométrie Algébrique* (Extracted from Bourbaki Seminar 1957-1962), Secrétariat mathématique, 11 rue Pierre Curie, PARIS 5e, 1962.

[AG3] A. Grothendieck and J.A. Dieudonné, *Eléments de Géométrie Algébrique,* Vol. I-IV, IHES, 1960-1967.

[AG4] A. Grothendieck and J.A. Dieudonné, *Eléments de Géométrie Algébrique I,* Springer-Verlag, 1970.

[AG5] A. Grothendieck, *Séminaire de Géométrie Algébrique,* Vol. 1-7, IHES, 1960-1962.

[HR] G. H. Hardy and S. Ramanujan, *Asymptotic Formulae in Combinatory Analysis,* Proc. London Math. Soc. (2), Vol. 17 (1917), pp. 75-115. (Reprinted in

Collected Papers of S. Ramanujan, pp. 276-309, Cambridge University Press, London, 1927.)

[H] G.H. Hardy, *A Mathematician's Apology*, Cambridge University. Press, 1969.

[GK1] G. Kempf, *Algebraic Varieties*, London Math. Soc. LNS 172, Cambridge Univ. Press, 1993.

[GK2] G. Kempf, *Algebraic Structures,* Friedrich Vieweg & Sohn, 1995.

[DM] D. Mumford, *Algebraic Geometry I, Complex Projective Varieties,* Springer-Verlag, 1976.

[JS] J. P. Serre, *Faisceaux Algébriques Cohérents*, Annals of Mathematics, Vol 61, No. 2, March, 1955.

[STS1] S.T. Soh, *A Reliable Implementation of the Formula of Rademacher for the Partition Function* $p(n)$, preprint

[STS2] S.T. Soh, *A Diophantine Construction of an Exact Algebraic Formula for Graded Partition Functions*, J. Korean Math. Soc., Vol 36 (1999), No. 2, pp. 267-298

[STS3] S.T. Soh, *Efficient (Quasi-) Recursive Formulas for Graded Partition Functions,* preprint

[ES] E.H. Spanier, *Algebraic Topology,* Springer-Verlag, 1966.

[BS] B. Sturmfels, *Algorithms in Invariant Theory,* Texts and Monographs in Symbolic Computation, Springer-Verlag, 1993.

[BW] B.L. van der Waerden, *Algebra,* Vol. 1 and 2, Frederick Ungar Pub., 1970.

[WR] A.N. Whitehead and B. Russell, *Principia Mathematica*, Cambridge Univ. Press, 1910.

[OZ] O. Zariski, *Collected Papers,* Vol. 1 - 4., MIT Press, 1972.

[ZP] O. Zariski and P. Samuel, *Commutative Algebra,* Vol.1 and 2, Springer-Verlag, 1958.

APPENDIX B

Computer Algebra Systems

In recent years, several Computer Algebra Systems were developed and they are currently used by many scientists and engineers. Among them, MAPLE, Mathematica, REDUCE, Macsyma, Matlab etc., are commercially available. In this appendix we first discuss several of MAPLE commands which are for Linear Algebra. Since this book is not written for the Computar Algebra System, MAPLE, our description of MAPLE commands is very brief. In the next section, we will discuss very briefly packages in REDUCE for Linear Algebra.

1. The Linear Algebra Package in MAPLE

To bring a package for Linear Algebra, type in at MAPLE prompt,

with(linalg); (and then press Enter key)

Then you are ready to use MAPLE's linear algebra package, linalg. In MAPLE, vectors and matrices are treated as the same kinds.

1.1. MAPLE commands for Matrices. Once you type in with(linalg); you will see a list of MAPLE commands available from the linalg package, and here are the list:

> with(linalg);

Warning, the protected names norm and trace have been redefined and unprotected

[BlockDiagonal, GramSchmidt, JordanBlock, LUdecomp, QRdecomp,
Wronskian, addcol, addrow, adj, adjoint, angle, augment,
backsub, band, basis, bezout, blockmatrix, charmat, charpoly,
cholesky, col, coldim, colspace, colspan, companion, concat,

cond, copyinto, crossprod, curl, definite, delcols, delrows, det, diag, diverge, dotprod, eigenvals, eigenvalues, eigenvectors, eigenvects, entermatrix, equal, exponential, extend, ffgausselim, fibonacci, forwardsub, frobenius, gausselim, gaussjord, geneqns, genmatrix, grad, hadamard, hermite, hessian, hilbert, htranspose, ihermite, indexfunc, innerprod, intbasis, inverse, ismith, issimilar, iszero, jacobian, jordan, kernel, laplacian, leastsqrs, linsolve, matadd, matrix, minor, minpoly, mulcol, mulrow, multiply, norm, normalize, nullspace, orthog, permanent, pivot, potential, randmatrix, randvector, rank, ratform, row, rowdim, rowspace, rowspan, rref, scalarmul, singularvals, smith, stackmatrix, submatrix, subvector, sumbasis, swapcol, swaprow, sylvester, toeplitz, trace, transpose, vandermonde, vecpotent, vectdim, vector, wronskian]

Among these commands, we explain several of them very briefly:

(1) matrix commands allows you to input a matrix,
(2) addrow, swaprow, and mulrow commands are for elementary row operations,
(3) inverse command computes the inverse of a matrix,
(4) det command computes the determinant of a matrix,
(5) &* is for matrix multiplication,
(6) evalm is for matrix evaluation,
(7) eigenvals, eigenvalues, eigenvectors, and eigenvects commands are for computation of eigenvectors and eigenvalues,
(8) nullspace command is for the nullspace of a matrix,
(9) colspace command is for the column space of a matrix,
(10) row space command is for the row space of a matrix,
(11) gausselim command is for the gaussian elimination of a matrix, etc.

1.2. Creating matrices.

EXAMPLE B.1. *Here are some examples:*
with(linalg);
A:=matrix([[1,2],[3,4]]);

$$A := \begin{bmatrix} 1 & 2 \\ 3 & 4 \end{bmatrix}$$

inverse(A),det(A);

$$\begin{bmatrix} -2 & 1 \\ 3/2 & -1/2 \end{bmatrix}, -2$$

B:=matrix(2,2,[w,x,y,z]):evalm(a+b);

$$\begin{bmatrix} 1+w & x+2x \\ 3+y & 4+z \end{bmatrix}$$

*evalm(A&*B);*

$$\begin{bmatrix} w+2y & x+2z \\ 3w+4y & 3x+4z \end{bmatrix}.$$

1.3. Accessing matrix components.

EXAMPLE B.2. *Here are some examples:*
with(linalg);
M1:=matrix([1,2,3,4],[a,b,c,d],[e,f,g,h],[w,x,y,z]]);

$$M1 := \begin{bmatrix} 1 & 2 & 3 & 4 \\ a & b & c & d \\ e & f & g & h \\ w & x & y & z \end{bmatrix}$$

M1[2,3];

$$c$$

submatrix(M1,1..2,3..4);

$$\begin{bmatrix} 3 & 4 \\ c & d \end{bmatrix}$$

subvector(M1,2,1..4);

$$[a, b, c, d]$$

row(M1,4);

$$[w, x, y, z]$$

col(M1,2);

$$[2, b, f, x]$$

1.4. Elementary arithmetic operations with matrices.

EXAMPLE B.3. *Here are some examples:*
with(linalg);
M1:=matrix([[1,2],[3,4]]): transpose(M1);

$$\begin{bmatrix} 1 & 2 \\ 3 & 4 \end{bmatrix}, \begin{bmatrix} 1 & 3 \\ 2 & 4 \end{bmatrix}$$

evalm(1/M1);

$$\begin{bmatrix} -2 & 1 \\ 3/2 & -1/2 \end{bmatrix}$$

map(sqrt, M1);

$$\begin{bmatrix} 1 & \sqrt{2} \\ \sqrt{3} & 2 \end{bmatrix}$$

*evalm(M1^3), eval(M1&*M!&*M1);*

$$\begin{bmatrix} 37 & 54 \\ 81 & 118 \end{bmatrix}, \begin{bmatrix} 37 & 54 \\ 81 & 118 \end{bmatrix}$$

1.5. Standard Examples in Linear Algebra.

EXAMPLE B.4. *Here are some examples:*
 # The following line invokes each of the inputs read in from a file to be displayed.
interface(echo=2);
with(plots):
with(plottools):
with(linalg):
*a:=matrix([[1,0,0,0,1,0,0,0],[5,2,0,0,0,1,0,0], *
[8,6,3,0,0,0,1,0],[0,9,7,4,0,0,0,1]]);

```
a1:=addrow(a,1,2,-5);
a2:=addrow(a1,1,3,-8);
a3:=addrow(a2,2,3,-3);
a4:=addrow(a3,2,4,-9/2);
a5:=mulrow(a4,2,1/2);
a6:=mulrow(a5,3,1/3);
a7:=addrow(a6,3,4,-7);
a8:=mulrow(a7,4,1/4);
aa:=submatrix(a,1..4,1..4);
b:=submatrix(a8,1..4,5..8);
check:=evalm(aa &* b);
b1:=vector([2,3,4,5]);
result:=linsolve(aa,b1);
check2:=evalm(aa &* result);
# The following line selects different plotdevice other than the default one.
# Using ghostview command at unix % prompt, you can display the graph of sin(x).
# interface( plotdevice=x11, plotoutput=myfile );
interface( plotdevice=ps, plotoutput=myfile );
p1:=plot3d(sin(x)*cos(y),x=-2*Pi..2*Pi,y=-2*Pi..2*Pi):
p2:=plot3d(x*y,x=-2*Pi..2*Pi,y=-2*Pi..2*Pi):
display({p1,p2});
# For instance, ghostview myfile (enter) at unix % prompt will display the graph of sin(x).
interface(plotdevice=default);
```

EXAMPLE B.5. *Here are some other examples:*

```
# The next line allows maple to disply what you typed in:
interface(echo=2);
with(linalg);
m1:=matrix([[a,b,c],[d,e,f],[g,h,i]]);
v1:=vector([u,v,w]);
```

```
v2:=matrix([[u],[v],[w]]);
evalm(m1 &*v1);
innerprod(v1,v1);
innerprod(v1,m1);
A:=matrix([[1,2,3],[4,8,12],[1,-1,0]]);
b:=vector([14,12,46]);
linsolve(A,b);
kernel(A);
nullspace(A);
eigenvals(A);
eigenvects(A);
poly:=charpoly(A,'x');
solve(poly,x);
v1:=vector([-1,2,3,0]);
v2:=vector([0,1,2,1]);
# v3:=vector([2,-1,-1,1]);
v3:=evalm(2*v2);
basis( {v1, v2, v3} );
GramSchmidt([v1,v2,v3],normalized);
```

1.6. How to Read MAPLE Codes. MAPLE is a procedural programming language, like C. With the exception of the commands belonging to the kernel and the compiled library commands, we can view the code of all MAPLE commands with 'print'.

EXAMPLE B.6. *Here are some examples:*

```
#To review the MAPLE command 'det' in Linear Algebra package,
with(linalg):
interface(verboseproc=2);
A:=mat([[1,2],[2,1]]);
det(A);
print(det);
# To review the MAPLE command 'charpoly' in Linear Algebra package,
```

with(linalg):
interface(verboseproc=2);
A:=mat([[1,2],[2,1]]);
charpoly(A,x);
print(charpoly);

2. The Linear Algebra Package in REDUCE

REDUCE is another well-known commercially available Computer Algebra System(CAS). To load a package for Linear Algebra, one needs to type in the following line at REDUCE prompt:

load_package linalg; (and then press Enter key)

This package provides the following commands, the usage of most of them are almost the same as those in MAPLE. The functions of the package can be divided into four categories:

2.1. Basic matrix handling. add_columns , add_rows , add_to_columns , add_to_rows , augment_columns , char_poly , column_dim , copy_into , diagonal , extend , find_companion , get_columns , get_rows , hermitian_tp , matrix_augment , matrix_stack , minor , mult_columns , mult_rows , pivot , remove_columns , remove_rows , row_dim , rows_pivot , stack_rows , sub_matrix , swap_columns , swap_entries , swap_rows .

Constructors – functions that create matrices

band_matrix , block_matrix , char_matrix , coeff_matrix , companion , hessian , hilbert , jacobian , jordan_block , make_identity , random_matrix , toeplitz , vandermonde .

2.2. High level algorithms. char_poly , cholesky , gram_schmidt , lu_decom , pseudo_inverse , simplex, svd .

2.3. Normal Forms. There is a separate package, NORMFORM, for computing the following matrix normal forms in REDUCE:

smithex , smithex_int , frobenius , ratjordan , jordansymbolic , jordan
.

2.4. Predicates. matrixp , squarep , symmetricp .

REMARK B.1. *For detailed usage informations on these commands, one may visit the following homepage:*

http://trinitas.mju.ac.kr/intro2reduce.html ,

although some part of the homepage is written in Korean, where one will easily find several manuals for REDUCE.[1]

[1]http://ch.catholic.or.kr/pundang/archive/4/intro2reduce.html

APPENDIX C

Suggestions for Further Reading and References

H. Anton and C. Rorres, *Elementary Linear Algebra*, 8th Edition, Wiley, 2000.

M. Artin, *Algebra*, Prentice Hall, 1991.

G. Birkhoff and S. MacLane, *A Survey of Modern Algebra*, 4th Edition, Macmillan, New Nork, 1997.

A.C. Hearn et al, *REDUCE User's Contributed Package Manual*, Version 3.8, Konrad-Zuse-Zentrum, Berlin, 2004.

G. Kempf, *Algebraic Structures,* Friedrich Vieweg & Sohn, 1995.

O.T. Kim and S.A. Park, *Linear Algebra* (Korean), 2nd Edition, Cheongmungak, 1995.

M. Kofler, *MAPLE: An Introduction and Reference*, Addison-Wesley, 1997.

J.H. Kwak, S.P. Hong, *Linear Algebra*, Birkhauser, 1997.

S. MacLane and G. Birkhoff, *Algebra*, 2nd Edition, Macmillan, New York, 1979.

S.T. Soh, *Computational Algebra*, Kyowoo, Seoul, 2006

S.T. Soh, *Algebraic Analysis I*, Kywoo, Seoul, 2006

S.T. Soh, *Abstract Algebra*, Kywoo, Seoul, 2007, 2019 (2nd Edition)

B.L. van der Waerden, *Modern Algebra*, Ungar, New York, 1970.

http://trinitas.mju.ac.kr/intro2softwares.html

http://trinitas.mju.ac.kr/intro2InetCompu.html

http://ch.catholic.or.kr/pundang/archive/4/soh/2002_InetCompu

http://www.reduce-algebra.com/documentation.php

http://www.reduce-algebra.com/manual/manual.pdf

http://www.reduce-algebra.com/manual/contributed/groebner.pdf

Index